ADVANCES IN COMPUTATIONAL STABILITY ANALYSIS

Edited by **Safa Bozkurt Coşkun**

Advances in Computational Stability Analysis
http://dx.doi.org/10.5772/3085
Edited by Safa Bozkurt Coşkun

Contributors

Richard Degenhardt, Alexander Kling, Rolf Zimmermann, Falk Odermann and F.C. de Araújo, Karam Maalawi, Jen-San Chen and Wei-Chia Ro, Huu-Tai Thai, Seval Pinarbasi Cuhadaroglu, Erkan Akpinar, Fuad Okay, Hilal Meydanli Atalay and Sevket Ozden, Safa Bozkurt Coşkun and Baki Öztürk

Published by InTech

Janeza Trdine 9, 51000 Rijeka, Croatia

Publishing Process Manager Marina Jozipovic
Technical Editor InTech Prepress, Novi Sad
Cover InTech Design team

Additional hard copies can be obtained from orders@intechopen.com

Advances in Computational Stability Analysis, Edited by Safa Bozkurt Coşkun
p. cm.
ISBN 978-953-51-0673-9

For EU product safety concerns:
IN TECH d.o.o., Prolaz Marije Krucifikse Kozulić 3, 51000 Rijeka, Croatia,
info@intechopen.com or visit our website at intechopen.com.

Meet the editor

Dr Safa Bozkurt Coşkun got both his BSc and his MSc degrees in Civil Engineering from the Middle East Technical University, Ankara, Turkey. He received his PhD in the field of Engineering Sciences in 2003 from the same University. He worked in the Department of Civil Engineering in Niğde University between 2004 and 2009. He has been working in the Department of Civil Engineering in Kocaeli University since 2009. Currently, he is an Associate Professor of Civil Engineering in the same University. His research interests include applied mathematics in engineering, vibration and stability theory, heat transfer, multiphase flow in porous media, fluid-structure interaction problems and finite and boundary element analyses. His researches have appeared in scientific journals such as Engineering Structures, Journal of Fluids and Structures, Applied Thermal Engineering, Computers and Mathematics with Applications, Engineering Computations, Structural Engineering and Mechanics, Mathematical Problems in Engineering and Mathematical Modeling and Analysis. He is also the co-author of a chapter in the book entitled: "Advances in Vibration Analysis Research" which was published by InTech in 2011.

Contents

Preface

Stability is a basic concern in both design and analysis of load-carrying systems and constitutes a major topic in the field of engineering science and mechanics. Since structural instability may lead to catastrophic failure of engineering structures, stability requirements must be satisfied besides requirements related to material failure. Knowledge on stability is of great importance in the areas of Civil Engineering, Mechanical Engineering and Aerospace Engineering; and all these disciplines have their own literature related to the subject.

This book is intended to present state-of-the art in the stability analysis and to bring a number of researches together exposing the advances in the field. It consists of original and innovative research studies exhibiting various investigation directions.

The advantage of this book is its being openly accessible and fully searchable. It provides a forum for dissemination of the most recent scientific information in the field of stability analysis. The subjects included will help the readers to learn more in this interesting field of study.

The audience of the book will include scientists in civil engineering, mechanical engineering, aerospace engineering and applied mathematics, graduate students as well as practicing engineers in the field.

I hope that the readers of this book will find it appealing and inspiring since it covers the most recent developments and innovations in the stability analysis field.

Safa Bozkurt Coşkun
Civil Engineering Department
Kocaeli University,
Turkey

Chapter 1

Dealing with Imperfection Sensitivity of Composite Structures Prone to Buckling

Richard Degenhardt, Alexander Kling,
Rolf Zimmermann, Falk Odermann and F.C. de Araújo

Additional information is available at the end of the chapter

http://dx.doi.org/10.5772/45810

1. Introduction

Currently, imperfection sensitive shell structures prone to buckling are designed according the NASA SP 8007 guideline using the conservative lower bound curve. This guideline dates from 1968, and the structural behaviour of composite material is not appropriately considered, in particular since the imperfection sensitivity and the buckling load of shells made of such materials depend on the lay-up design. This is not considered in the NASA SP 8007, which allows designing only so called "black metal" structures. There is a high need for a new precise and fast design approach for imperfection sensitive composite structures which allows significant reduction of structural weight and design cost. For that purpose a combined methodology from the Single Perturbation Load Approach (SPLA) and a specific stochastic approach is proposed which guarantees an effective and robust design. The SPLA is based on the observation, that a large enough disturbing load leads to the worst imperfection; it deals with the traditional (geometric and loading) imperfections [1]. The stochastic approach considers the non-traditional ones, e.g. variations of wall thickness and stiffness. Thus the combined approach copes with both types of imperfections. A recent investigation demonstrated, that applying this methodology to an axially loaded unstiffened cylinder is leading directly to the design buckling load 45% higher compared with the respective NASA SP 8007 design [2].

This chapter presents in its first part the state-of-the-art in buckling of imperfection sensitive composite shells. The second part describes current investigations as to the SPLA, the stochastic approach and their combination. In a third part an outlook is given on further studies on this topic, which will be performed within the framework of the running 3-year project DESICOS (New Robust DESIgn Guideline for Imperfection Sensitive COmposite Launcher Structures) funded by the European Commission; for most relevant architectures of cylindrical and conical launcher structures (monolithic, sandwich - without and with holes) the new methodology will be further developed, validated by tests and summarized in a handbook for the design of imperfection sensitive composite structures. The potential will be demonstrated within different industrially driven use cases.

2. State of the art

2.1. Imperfection sensitivity

In Figure 1 taken from [3], knock-down factors – the relations of experimentally found buckling loads and of those computed by application of the classical buckling theory - are shown for axially compressed cylindrical shells depending on the slenderness. The results are presented by dots and show the large scatter. The knock-down factors decrease with increasing slenderness. The discrepancy between test and classical buckling theory has stimulated scientists and engineers on this subject during the past 50 years. The efforts focused on postbuckling, load-deflection behaviour of perfect shells, various boundary conditions and their effect on bifurcation buckling, empirically derived design formulas and initial geometric imperfections. Koiter was the first to develop a theory which provides the most rational explanation of the large discrepancy between test and theory for the buckling of axially compressed cylindrical shells. In his doctoral thesis published in 1945 Koiter revealed the extreme sensitivity of buckling loads to initial geometric imperfections. His work received little attention until the early 1960's, because the thesis was written in Dutch. An English translation by Riks was published 1967 in [4].

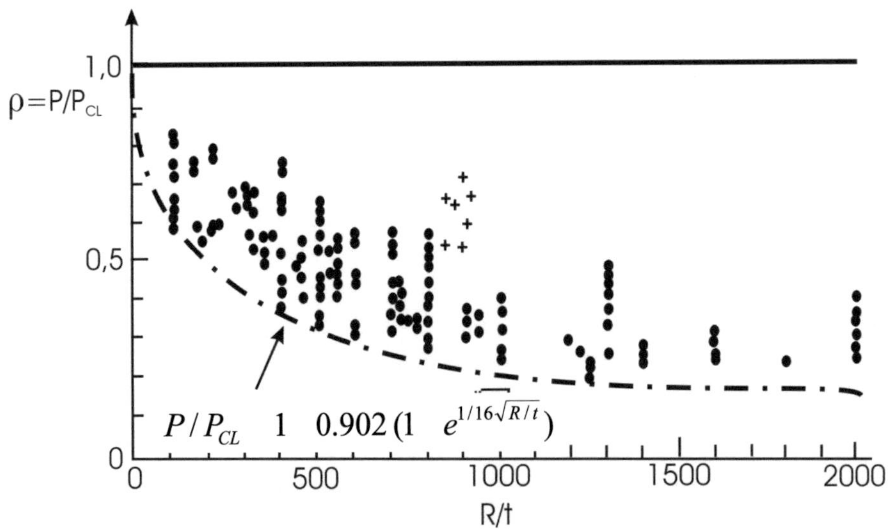

Figure 1. Distribution of test data for cylinders subjected to axial compression [1]

Based on a number of experimental tests in the 1950s and 60s the determination of lower bounds led to design regulations like NASA SP-8007 [1], but the given knock-down factors are very conservative. To improve the ratio of weight and stiffness and to reduce time and cost, numerical simulations could be used during the design process. The consideration of imperfections in the numerical simulation is essential for safe constructions. Usually, these imperfections are unknown in the design phase, thus pattern and amplitude have to be assumed.

In general, one can distinguish between loading imperfections and geometric imperfections. Both kinds of imperfections have a significant influence on the buckling behaviour.

Loading imperfections mean any deviations from perfect uniformly distributed loading, independent of the reason of the perturbation. Geier et al. tested composite cylindrical shells with different laminate designs [5], and they applied thin metal plates locally between test shell and supporting structure to perturb the applied loads and performed the so-called shim tests [6]. Later, numerical investigations were performed and compared to the test results; the importance was verified [7]. The need to investigate loading imperfections for practical use was shown for instance by Albus et al. [8] by the example of Ariane 5.

Geometric imperfections mean any deviations from the ideal shape of the shell structure. They are often regarded the main source for the differences between computed and tested buckling loads. Winterstetter et al. [9] suggest three approaches for the numerical simulation of geometrically imperfect shell structures: "realistic", "worst" and "stimulating" geometric imperfections. Stimulating geometric imperfections like welded seams are local perturbations which "stimulate" the characteristic physical shell buckling behaviour [10]. "Worst" geometric imperfections have a mathematically determined worst possible imperfection pattern like the single buckle [11]. "Realistic" geometric imperfections are determined by measurement after fabrication and installation. This concept of measured imperfections is initiated and intensively promoted by Arbocz [12]; a large number of test data is needed, which has to be classified and analysed in an imperfection data bank. Within the study presented in this paper, real geometric imperfections measured at test shells are taken into account.

Hühne et al. [1] showed that for both, loading imperfections and geometric imperfections the loss of stability is initiated by a local single buckle. Therefore unification of imperfection sensitivity is allowed; systems sensitive to geometric imperfections are also sensitive to loading imperfections. Single buckles are realistic, stimulating and worst geometric imperfections.

Using laminated composites, the structural behaviour can be tailored by variation of fibre orientations, layer thicknesses and stacking sequence. Fixing the layer thicknesses and the number of layers, Zimmermann [13] demonstrated numerically and experimentally that variation of fibre orientations affects the buckling load remarkably. The tests showed that fibre orientations can also significantly influence the sensitivity of cylindrical shells to imperfections. Meyer-Piening et al. [14] reported about testing of composite cylinders, including combined axial and torsion loading, and compared the results with computations.

Hühne [1] selected some of the tests described in [13] to [15] and performed additional studies. Within a DLR-ESA study one of these cylinder designs, which is most imperfection sensitive, was manufactured 10 times and tested. It allowed a comparison with already available results and enlarged the data base [2].

2.2. Single-perturbation-load approach

Hühne [1] proposed an approach based on a single buckle as the worst imperfection mode leading directly to the load carrying capacity of a cylinder. Figure 2 explains its mechanism; the lateral perturbation load P is disturbing the otherwise unloaded shell, and the axial compression load F is applied until buckling. This is repeated with a series of different perturbation loads, starting with the undisturbed shell and the respective buckling load F_0. In Figure 3 buckling loads F depending on the perturbation loads P are depicted. The figure shows that the buckling load belonging to a perturbation load larger than a minimum value P_1 is almost constant. A further increase of the pertubation load has no significant change on the buckling any more. The buckling load F_1 is considered to be the design buckling load. This concept promises to improve the knock-down factors and allows designing any CFRP cylinder by means of one calculation under axial compression and a single-perturbation load. Within a DLR-ESA study, this approach was confirmed analytically and experimentally, cf. [2]. However, there is still the need for a multitude of further studies.

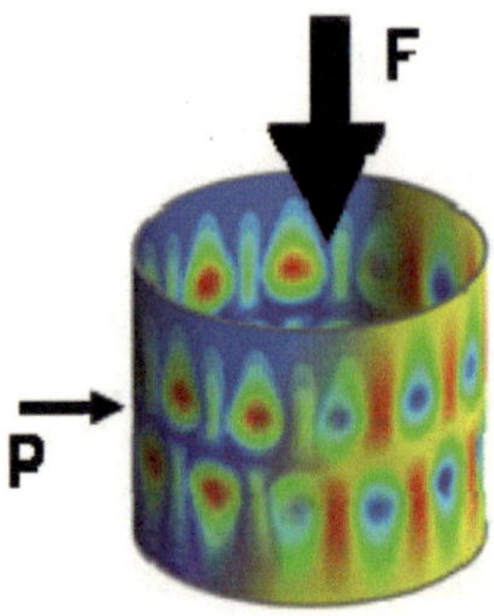

Figure 2. Perturbation load mechanism

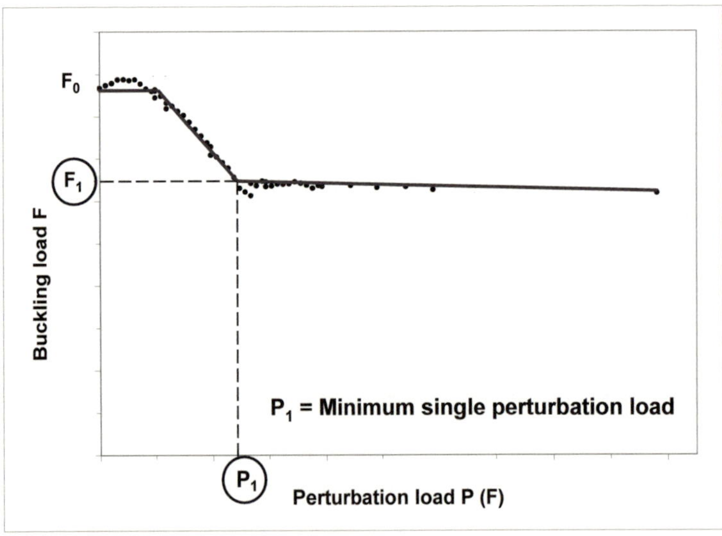

Figure 3. Single perturbation load approach (SPLA)

2.3. Probabilistic research

In general, tests or analysis results are sensitive to certain parameters as boundary conditions or imperfections. Probabilistic methods are a possibility to assess the quality of results. The stochastic simulation with Monte Carlo (e.g. [17]) allows the statistical description of the sensitivity of the structural behaviour. It starts with a nominal model and makes copies of it whereas certain parameters are varied randomly. The random numbers, however, follow a given statistical distribution. Each generated model is slightly different, as in reality.

Recently, probabilistic simulations found the way into all industrial fields. In automotive engineering they are successfully applied in crash or safety (e.g. [18]). Klein et al. [20] applied the probabilistic approach to structural factors of safety in aerospace. Sickinger and Herbeck [21] investigated the deployable CFRP booms for a solar propelled sail of a spacecraft using the Monte Carlo method.

Velds [22] performed deterministic and probabilistic investigations on isotropic cylindrical shells applying finite element buckling analyses and showed the possibility to improve the knock-down factors. However, setting-up of a probabilistic design approach still suffers by a lack of knowledge due to the incomplete base of material properties, geometric deviations, etc..

Arbocz and Hilburger [23] published a probability-based analysis method for predicting buckling loads of axially compressed composite cylinders. This method, which is based on the Monte Carlo method and first-order second-moment method, can be used to form the basis for a design approach and shell analysis that includes the effects of initial geometric imperfections on the buckling load of the shell. This promising approach yields less conservative knock-down factors than those used presently by industry.

2.4. Specific stochastic approach

Figure 4 shows the variation (gray shaded band) of the buckling load resulting from its sensitivity to the scatter of the non-traditional imperfections (e.g. thickness variations). It demonstrates the need to cover this by the development of an additional knock-down factor ρ_2 in combination to the knock-down factor ρ_1 from SPLA.

An efficient design is feasible, if knowledge about possibly occurring imperfections exists and if this knowledge is used within the design process. Whereas the traditional imperfections are dealt with the SPLA, the non-traditional ones are taken into account by probabilistic methods, which enable the prediction of a stochastic distribution of buckling loads. Once the distribution of buckling loads is known, a lower bound can be defined by choosing a level of reliability. Degenhardt et al. [2] found less conservative knockdown factors than through the NASA-SP 8007 lower bound, by executing probabilistic analyses with non-traditional imperfections.

The work for the stochastic approach consists in checking which structural parameters substantially influence the buckling load and defining realistic limits for their deviations from the nominal values, in varying them within the limits and performing buckling load computations for these variations. The results are evaluated stochastically in order to define a guideline for the lower limits of the buckling loads within a certain given reliability. From these limits a knock-down factor is derived.

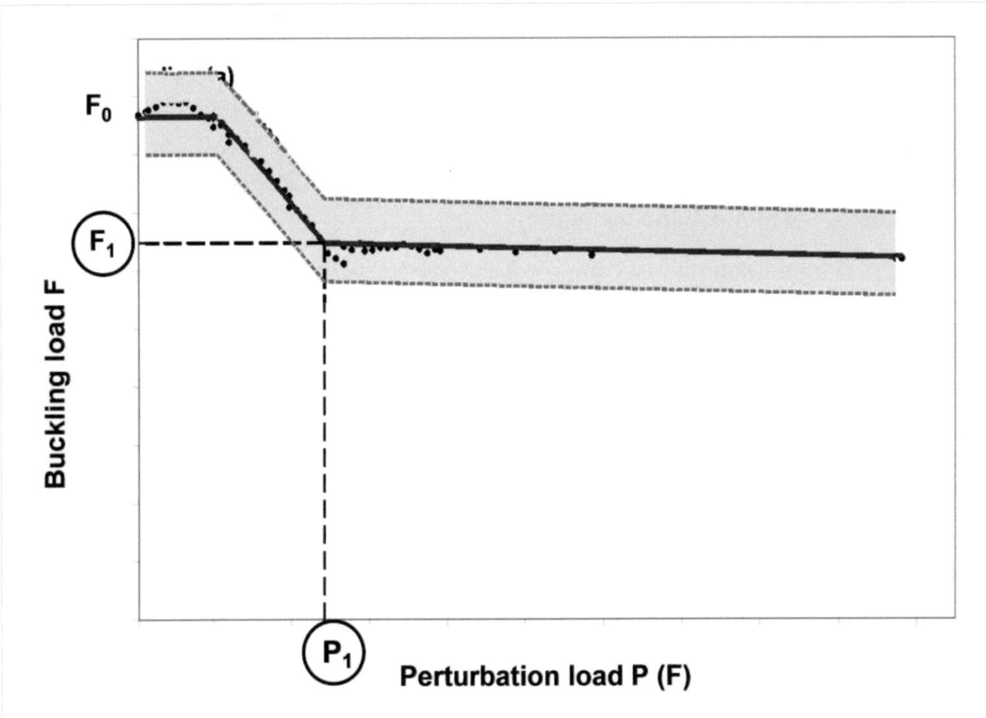

Figure 4. Scatter of buckling load due to the scatter of non-traditional imperfections

2.5. Conclusions

From all this it becomes obvious that a great deal of knowledge is accumulated concerning the buckling of cylindrical shells under axial compression. However, the NASA guideline for the knock-down factors from 1968 is still in use, and there are no appropriate guidelines for unstiffened cylindrical CFRP shells. To define a lower bound of the buckling load of CFRP structures a new guideline is needed which takes the lay-up and the imperfections into account. This can be for instance a probabilistic approach or the Single-Perturbation-Load approach, combined with a specific stochastic approach. In the following the second one is considered in more detail. Independent of the approach dozens of additional tests are necessary, in order to account for statistical scatter as well as for software and guideline validation

3. SPLA combined with specific stochastic approach

3.1. The procedure and first results

Figure 5 summarises the future design scenario for imperfection sensitive composite structures in comparison to the current design scenario. Currently, the buckling load of the perfect structure $F_{Perfect}$ has to be multiplied by the knock-down factor $_{NASA}$ from the NASA SP 8007 guideline. This approach was developed for metallic structures in 1968 and does not at all allow exploiting the capacities of composite structures. Accordingly, with the new design scenario $F_{Perfect}$ is multiplied by ρ_1 which results from SPLA and ρ_2 which comes from the specific stochastic approach.

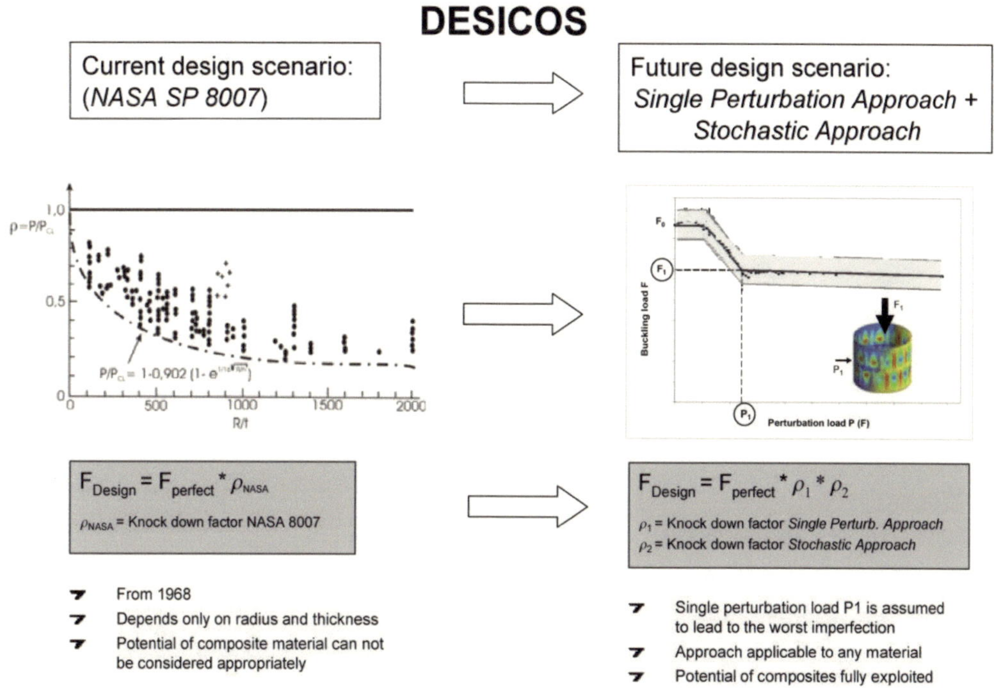

Figure 5. Future design scenario for composite structures

First studies (cf. [2]) demonstrated the high potential of this combined approach which is summarized in Figure 6. In this example a composite cylinder (R/t=500) with 4 layers was designed according the current and the future design scenarios. The classical buckling load was calculated and utilized as reference (scaled buckling load ϱ=1.0, marked by a star). The buckling load calculated by the SPLA was at ϱ=0.58 (marked by a star). All experimentally extracted results revealed first buckling beyond the one calculated by the SPLA (safe design). The knock down factor from the SPLA was found to be 0.58 (times 0.8 from stochastic), whereas the one form NASA SP was 0.32. The result was that the load carrying capacity could be increased by 45%. It corresponds to approximately 20% weight reduction for the same load. In [2] the results were validated by tests on 10 nominally identical structures.

The improvement of load carrying capacity by 45% for the investigated 4-ply laminate can be considered to be representative for the following reasons: That laminate set-up was chosen because of its remarkable imperfection sensitivity known from foregoing investigations. With high imperfection sensitivity NASA SP 8007 is not as conservative as with a lower one, nevertheless the improvement of load carrying capacity came to 45%. With lay-ups leading to low imperfection sensitivity the NASA SP 8007 is extremely conservative because it is overestimating the negative influence of imperfections. In that case the improvement of load carrying capacity may be even higher than 45%. Thus the margin of 45% is at the lower limit of improvement of load carrying capacity, and it is not relevant for the expected improvement due to the novel approach whether the 4-ply laminate is optimal or representative for the real construction.

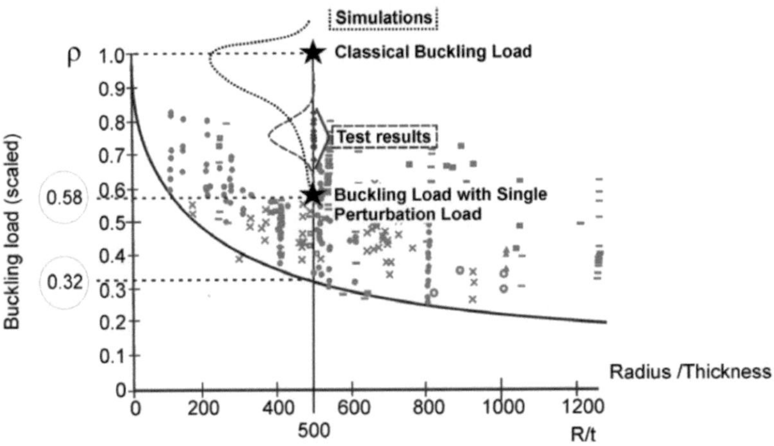

- $\rho_{NASA} = 0.32$
- $\rho_1 = 0.58$
- $\rho_2 = 0.80$ (Conservative estimation)

$F_{Design} = F_{perfect} * \rho_{NASA}$

$F_{Design} = 32 * 0.32 = \mathbf{10.2\ kN}$

$F_{Design} = F_{perfect} * \rho_1 * \rho_2$

$F_{Design} = 32 * 0.58 * 0.8 = \mathbf{14.8\ kN}$

Figure 6. Potential of the future design scenario [2]
Example: CFRP cylindrical shell (R/t = 500, 4 layers), $F_{perfect} = 32$ kN

3.2. The key role of experiments

New design methods or new software tools in the engineering have to be validated by test results. In addition, stochastic approaches require comprehensive data bases. In order to achieve suitable results appropriate test facilities and measurement systems, but also experience is needed. In the following, facilities and procedures are listed as currently used at DLR, cf. [16].

The buckling test facility is the main instrument to investigate buckling phenomena and to validate software simulations. Figure 7 shows on the left the axial compression configuration and on the right the compression-shear-configuration of the buckling test facility of the DLR Institute of Composite Structures and Adaptive Systems. The test facility can be changed from one configuration to another according to the test requirements.

Figure 7. DLR's buckling test facility, axial compression configuration (left), compression-shear-configuration (right)

The axial compression configuration is best suited for investigation of imperfection sensitivity on cylindrical structures. All parts of the test device are extremely stiff. The test specimen is located between an axially supporting top plate and a lower drive plate. The top plate can be moved in vertical direction on three spindle columns in order to adapt the test device to various lengths of test specimens. Due to the great sensitivity of stability tests against non-uniform load introduction even the small necessary clearance inside the spindle drives is fixed during the tests by automatically operating hydraulic clamps. The top plate functions as a counter bearing to the axial force that is applied to the movable lower drive plate by a servo-controlled hydraulic cylinder. The drive plate acts against the specimen, which itself acts against a stout cylindrical structure that is meant to distribute the three concentrated forces coming from three load cells at its upper surface, into a smooth force distribution. The test specimen is placed between the load distributor and the drive plate. Although the test device and test specimen are manufactured with particular care one can not expect, that the fixed upper plate and the load distributor are perfectly plane and parallel to each other, nor can one expect the end plates or clamping boxes of the test specimen to be perfectly plane and parallel. To make sure, that the test specimen will be uniformly loaded, thin layers of a kind of epoxy concrete, i.e. epoxy reinforced with a mixture of sand and quartz powder, are applied between the end plates or clamping boxes of the test specimens and the adjacent parts of the test device. This has the side effect of securing the test specimens against lateral displacement. In order to determine the offset of the load measurement it is required, that at least one side of the specimen may be separated temporarily from the test facility. This is achieved by using a separating foil between the top plate and the upper epoxy layer. Two displacement transducers (LVDT) are used to measure axial shortening of the specimen during the tests. Their signals are recorded and, moreover, used for control purposes as actual values. Hence, the test device is displacement controlled. According to the particular arrangement of the transducers the elastic deformation of the test device does not influence the control by shortening at quasi-static loading. Table 1 summarizes the characteristics of the test facility.

Load case	
Axial compression	Max. 1000 kN
Torsion	Max. 20 kNm
Internal pressure	Max. 800 kPa
External pressure	Max. 80 kPa
Shear	Max. 500 kN
Geometry limits of the test structure	
Length	Max. 2100 mm
Width (diameter)	Max. 1000 mm
Load frequency (axial compression only)	Max. 50 Hz

Table 1. Characteristics of the DLR buckling test facility

Before testing geometric and material imperfections are measured by the following systems (or equivalent)

1. ATOS: The ATOS system is based on photogrammetry (precision: 0.02 mm), Figure 8 illustrates an example of measured imperfections which are scaled by a factor of 100 to improve the visibility
2. Ultrasonic inspection

During testing the deformations are measured by the ARAMIS system:

It is based on photogrammetry (precision: 0.02 mm). The system used allows also a 360° measurement of shell surface displacements on a CFRP cylinder (cf. Figure 9 and Figure 11).

All measured full field displacements are transferred to a global coordinate system of the cylinder by means of at least three reference points in each area. The reference points are allocated to the global coordinate system by TRITOP, another photogrammetric system. The result of this procedure is a complete 3-D visualisation of the cylinder deformation (cf. Figure 9). The four camera pairs can also be placed on one part of the structure which allows a quadruplicating of the number of taken pictures per time (Figure 10). A 360° survey of a CFRP cylinder (selected deformation patterns of one loading and unloading sequence) is presented in Figure 11.

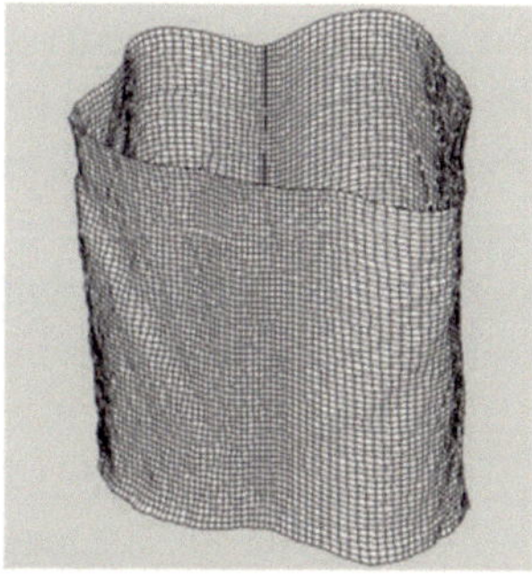

Figure 8. Measured geometric imperfections (ATOS)

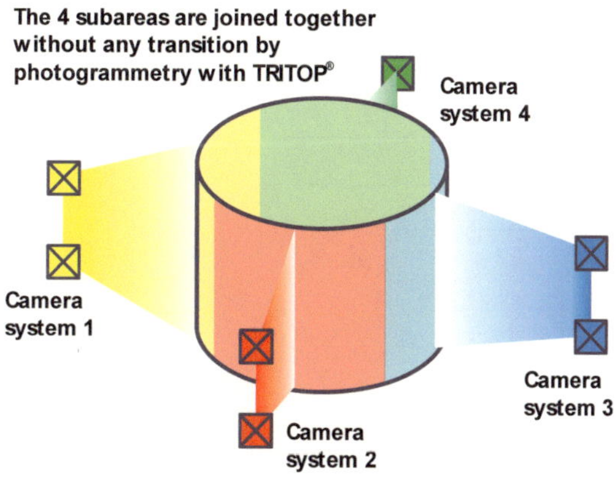

Figure 9. 360° Measurement on a cylinder

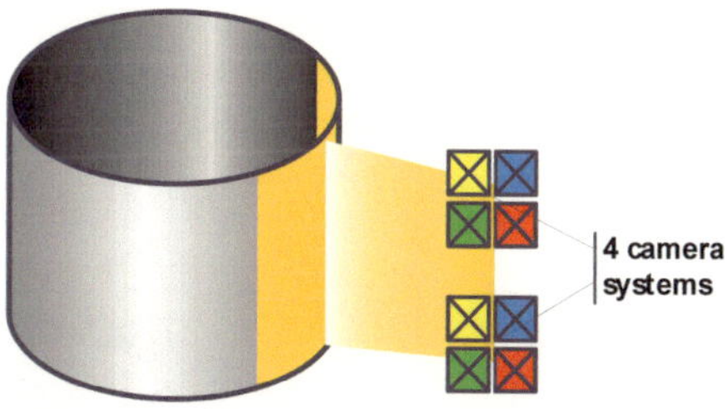

Figure 10. High Speed ARAMIS Set-Up

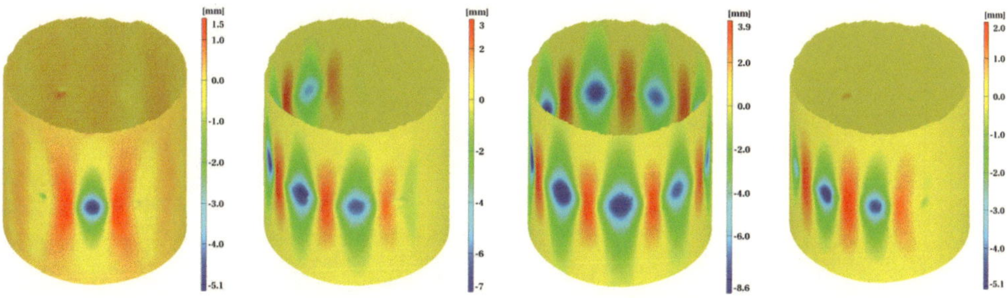

Figure 11. Results of a 360° Measurement on a CFRP cylinder (selected deformation patterns of one loading and unloading sequence)

Figure 12 illustrates the measured load-shortening curves of 10 tests with three selected pictures extracted from ARAMIS measurement obtained from the 360° measurement. Picture A and B are from the pre-buckling and Picture C from the early post-buckling state. Figure 13 compares the post-buckling pattern between test and Finite-Element simulation. The left picture is obtained by the 360° ARAMIS measurement. It agrees quite well with the simulation in the right figure. This buckling pattern was observed for all 10 cylinders. More details can be found in [2].

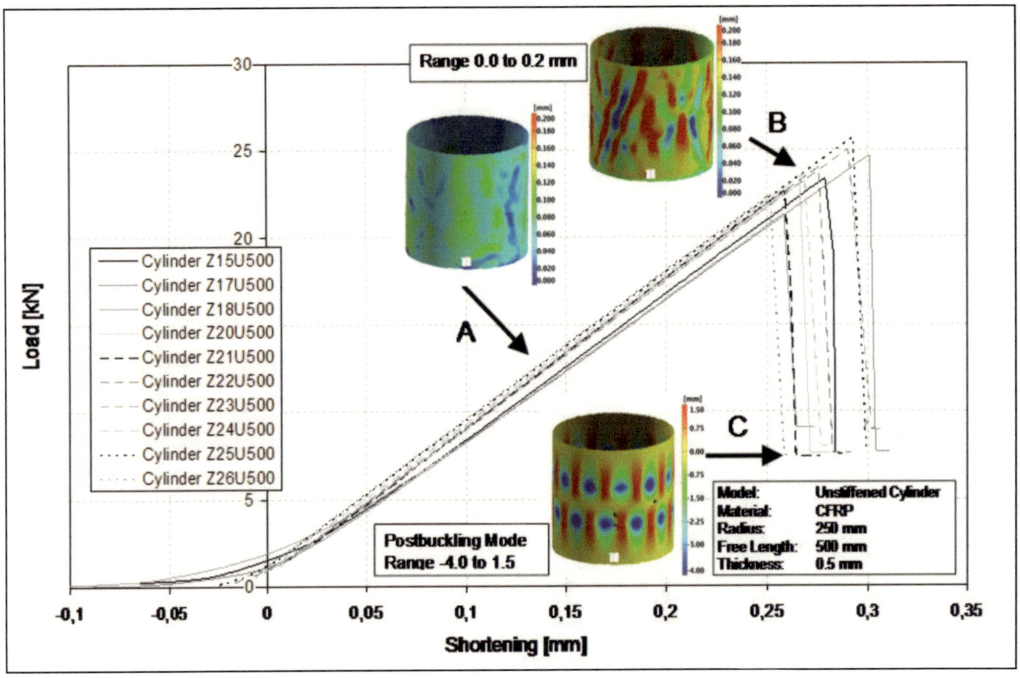

Figure 12. Load shortening curves of 10 tested cylinders and ARAMIS measurement

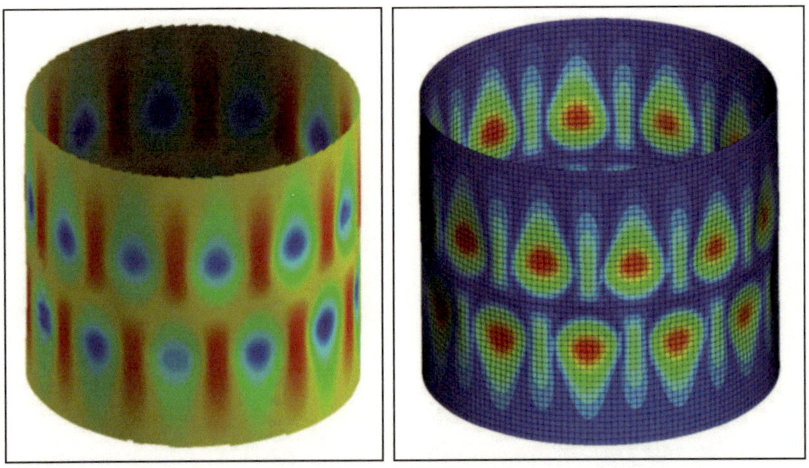

Figure 13. Postbuckling pattern. Left: Test results obtained by ARAMIS – Right: Simulation results

4. DESICOS project

4.1. Main objective

The main objective of DESICOS is to establish an approach on how to handle imperfection sensitivity in space structures endangered by buckling, in particular for those made from fiber composite materials. It shall substitute the NASA SP 8007, which is extremely conservative and not really applicable for composite structures, cf. Figure 5.

The DESICOS consortium merges knowledge from 2 large industrial partners (ADTRIUM-SAS from France and Astrium GmbH from Germany), one enterprise belonging to the category of SME (GRIPHUS from Israel), 2 research establishments (DLR from Germany and CRC-ACS from Australia) and 7 universities (Politecnico di Milano from Italy, RWTH Aachen, Leibniz University and the Private University of Applied Sciences Göttingen from Germany, TECHNION from Israel, TU-Delft from Netherlands and Technical University of Riga from Latvia). The large industrial enterprises and the SME bring in their specific experience with designing and manufacturing of space structures as well as their long grown manufacturing philosophies for high quality stiffened composite structures. The academic partners and the research organisations provide their special knowledge in methods and tool development as well as testing. This consortium composition assures the expected rapid and extensive industrial application of the DESICOS results.

4.2. Workpackages

The partners co-operate in the following technical work packages:

- **WP1:** Benchmarking on selected structures with existing methods: Benchmarks are defined for method evaluation purposes. The objective is the knowledge of the abilities and deficiencies of existing approaches.
- **WP2:** Material characterisation and design of structures for buckling tests: The first focus is on the design of structures which will be manufactured and tested in WP4. For that purpose, small specimens will be built and tested in order to characterise the specific composite material properties.
- **WP3:** Development and application of improved design approaches: In this workpackage new design approaches are developed, modelling and analysis strategies are derived. Finally, all methods are validated by means of the experimental results obtained from the other workpackages.
- **WP4:** Manufacture, inspection and testing of structures designed in WP 2: This workpackage deals with the manufacturing and testing of structures. The objective is to extend the data base on buckling of imperfection sensitive structures. Based on the designs from WP2 as input, a total of 14 (monolithic, sandwich, stiffened and unstiffened, cylindrical and conical) structures will be considered.
- **WP5:** Design handbook and industrial validation: WP5 comprises the final technical part; all the results of the project are assembled in order to derive the final design guidelines and to validate them as well as the new methods. The output is summarized

in the improved design procedures, the documentation of the designs as well as the documentation of the experiments and their evaluated results.

4.3. Expected results

To reach the main objective, improved design methods, experimental data bases as well as design guidelines for imperfection sensitive structures are needed. The experimental data bases are indispensable for validation of the analytically developed methods. Reliable fast methods will allow for an economic design process. Industry brings in experience with the design and manufacture of real shells; research contributes knowledge on testing and on development of design methods. Design guidelines are defined in common, and the developed methods are validated by industry.

The results of DESICOS comprise:

- Material properties, measured according to the applicable standards
- Method for the design of buckling critical fibre composite launcher structures, based on the combined SPLA and stochastic procedures, validated by experiments
- Experimental results of buckling tests including measured imperfections, buckling and postbuckling deformations, load shortening curves, buckling loads
- Guidelines how to design composite cylindrical shells to resist buckling
- Reliable procedure how to apply the Vibration Correlation Technique (VCT) in order to predict buckling loads non-destructively by experiments
- Handbook including all the results
- Demonstration of the potential with different industrially driven use cases.

5. Summary

This chapter summarises the state-of-the-art of imperfection sensitive composite structures prone to buckling. The current design process according the NASA SP 8007 is shown and its limitations to design structures made of composites are explained. A new promising approach which combines the Single Perturbation Load Approach and a Stochastic Approach - as an alternative to the NASA SP 8007 - is presented. It is further developed in the EU project DESICOS the objectives and expected results of which are given. More details can be found at www.desicos.de.

Author details

Richard Degenhardt*
DLR, Institute of Composite Structures and Adaptive Systems, Braunschweig, Germany
PFH, Private University of Applied Sciences Göttingen,
Composite Engineering Campus Stade, Germany

* Corresponding Author

Alexander Kling, Rolf Zimmermann and Falk Odermann
DLR, Institute of Composite Structures and Adaptive Systems, Braunschweig, Germany

F.C. de Araújo
Dept Civil Eng, UFOP, Ouro Preto, MG, Brazil

Acknowledgement

The research leading to these results has received funding from the European Community's Seventh Framework Programme (FP7/2007-2013) under Priority Space, Grant Agreement Number 282522. The information in this paper reflects only the author's views and the European Community is not liable for any use that may be made of the information contained therein.

6. References

[1] C. Hühne, R. Rolfes, E. Breitbach, J. Teßmer, „Robust Design of Composite Cylindrical Shells under Axial Compression — Simulation and Validation", *Thin-Walled Structures* Vol. 46 (2008) pp. 947–962

[2] Degenhardt R., Kling A., Bethge A., Orf J., Kärger L., Rohwer K., Zimmermann R., Calvi A., "Investigations on Imperfection Sensitivity and Deduction of Improved Knock-Down Factors for Unstiffened CFRP Cylindrical Shells", *Composite Structures* Vol. 92 (8), (2010), pp. 1939–1946

[3] P. Seide, V.I. Weingarten, J.P. Petersen, NASA/SP-8007, Buckling of thin-walled circular cylinders, *NASA SPACE VEHICLE* DESIGN *CRITERIA (Structures)*, NASA ,Washington, DC, United States, September, 1965

[4] W.T. Koiter, On the Stability of Elastic Equilibrium. NASA-TT-F-10833, 1967

[5] B. Geier, H. Klein and R. Zimmermann, Buckling tests with axially compressed unstiffened cylindrical shells made from CFRP, Proceedings, *Int. Colloquium on Buckling of shell Structures*, on land, in the sea and in the air, J.F. Julien, ed.: Elsevier Applied Sciences, London and New York, 498-507, 1991

[6] Geier, B., H. Klein and R. Zimmermann, Experiments on buckling of CFRP cylindrical shells under non-uniform axial load, Proceedings, *Int. Conference on Composite Engineering*, ICCE/1 28-31 August 1994

[7] C. Hühne, R. Zimmermann, R. Rolfes and B. Geier, Loading imperfections – Experiments and Computations, *Euromech Colloquium 424*, Kerkrade, The Netherlands, 2-5 September 2001

[8] J. Albus, J. Gomez-Garcia, H. Oery, Control of Assembly Induced Stresses and Deformations due to Waviness of the Interface Flanges of the ESC-A Upper Stage, *52nd International Astronautical Congress*, Toulouse, France, 1-5 Oct 2001

[9] T. Winterstetter, H. Schmidt, Stability of circular cylindrical steel shells under combined loading, *Thin-Walled Structures*, 40, p. 893-909, 2002

[10] M. Pircher, R. Bridge, The influence of circumferential weld-induced imperfections on the buckling of silos and tanks, *Journal of Constructional Steel Research*, 57(5), p. 569-580, 2001

[11] M. Deml, W. Wunderlich, Direct evaluation of the 'worst' imperfection shape in shell buckling, *Computer Methods in Applied Mechanics and Engineering*, 149 [1-4], p. 201-222, 1997

[12] J. Arbocz, Jr. J.H. Starnes, Future directions and challenges in shell stability analysis, *Thin-Walled Structures*, 40, p. 729-754, 2002

[13] R. Zimmermann, Buckling Research for Imperfection tolerant fibre composite structures, Proc. *Conference on Spacecraft Structures*, Material & Mechanical Testing, Noordwijk, The Netherlands, ESA SP-386, 27-29 March 1996

[14] H.-R. Meyer-Piening, M. Farshad, B. Geier, R. Zimmermann, Buckling loads of CFRP composite cylinders under combined axial and torsion loading - experiment and computations, *Composite Structures* 53, 427-435, 2001

[15] R. Zimmermann, Optimierung axial gedrückter CFK-Zylinderschalen, Fortschrittsberichte VDI, Nr. 207, 1992

[16] Degenhardt R., Kling A., Klein H., Hillger W., Goetting Ch., Zimmermann R., Rohwer K., Gleiter A., "Experiments on Buckling and Postbuckling of Thin-Walled CFRP Structures using Advanced Measurement Systems", *International Journal of Structural Stability and Dynamics*, Vol. 7, No. 2 (2007), pp. 337-358

[17] C.A. Schenk, G.I. Schuëller, Buckling Analysis of Cylindrical Shells with Cutouts including Random Geometric Imperfections, WCCM V, 7-12 July 2002

[18] R. Reuter, T. Gärtner, Stochastische Crashsimulation mit LS-DYNA am Beispiel des Kopfaufpralls nach FMVSS 201, *17 CAD-FEM Users meeting*, Sonhofen, 1999

[19] R. Reuter, J. Hülsmann, Achieving design targets through stochastic simulation, *8th Int. MADYMO Users Conference*, Paris, 2000

[20] M. Klein, G.I. Schuëller, P. Deymarie, M. Macke, P. Courrian, R. S. Capitanio, Probabilistic approach to structural factors of safety in aerospace, Proceedings of the *International Conference on Spacecraft Structures and Mechanical Testing*, pages 679–693, Cépadués-Editions, **2-210, Paris, France, 1994

[21] C. Sickinger, L. Herbeck, E. Breitbach, Structural engineering on deployable CFRP booms for a solar propelled sailcraft, *54th Int. Astronautical Congress*, Bremen, Germany, 28 Sep.-3 Oct., 2003

[22] E. Velds, TUD/LR/ASCM graduation thesis, A deterministic and probabilistic approach to finite element buckling analysis of imperfect cylindrical shells, May 2002

[23] J. Arbocz, M.W. Hilburger, Toward a Probabilistic Preliminary Design Criterion for Buckling Critical Composite Shells, *AIAA Journal, Volume 43,* No. 8, pp. 1823-1827, 2005

Chapter 2

Stability, Dynamic and Aeroelastic Optimization of Functionally Graded Composite Structures

Karam Maalawi

Additional information is available at the end of the chapter

http://dx.doi.org/10.5772/45878

1. Introduction

Numerous applications of numerical optimization to various structural design problems have been addressed in the literature. A comprehensive survey on this issue was given in [1], presenting a historical review and demonstrating the future needs to assimilate this technology into the practicing design environment. Different approaches were applied successfully by several investigators for treating stress, displacement, buckling and frequency optimization problems. In general, design optimization seeks the best values of a set of n design variables represented by the vector, $\underline{X}_{nx1}$, to achieve, within certain m constraints, $\underline{G}_{mx1}(\underline{X})$, its goal of optimality defined by a set of k objective functions, $\underline{F}_{kx1}(\underline{X})$, for specified environmental conditions (see Figure 1). Mathematically, design optimization may be cast in the following standard form: Find the design variables $\underline{X}_{nx1}$ that minimize

$$F(\underline{X}) = \sum_{i=1}^{k} W_{fi} F_i(\underline{X}) \tag{1a}$$

subject to

$$G_j(\underline{X}) \leq 0\ ,\ j=1,2,\dots\dots\dots I \tag{1b}$$

$$G_j(\underline{X}) = 0\ ,\ j=I+1,I+2,\dots.m \tag{1c}$$

$$0 \leq W_{fi} \leq 1 \qquad \sum_{i=1}^{k} W_{fi} = 1 \tag{1d}$$

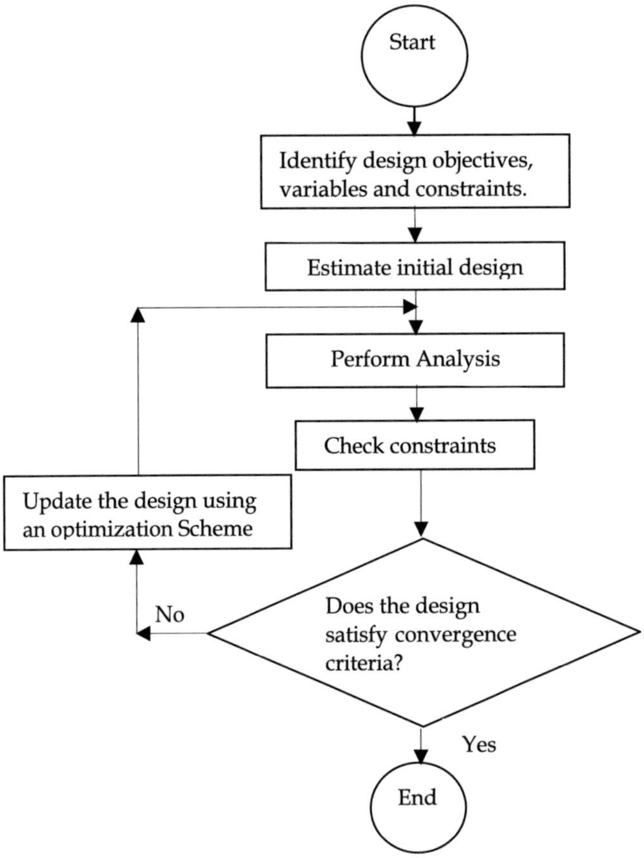

Figure 1. Design optimization process

The weighting factors W_{fi} measure the relative importance of the individual objectives with respect to the overall design goal. Several computer program packages are available now for solving a variety of design optimization models. Advanced procedures are carried out by using large-scale, general purpose, finite element-based multidisciplinary computer programs [2], such as *ASTROS, MSC/NASTRAN* and *ANSYS*. The *MATLAB* optimization toolbox is also a powerful tool that includes many routines for different types of optimization encompassing both unconstrained and constrained minimization algorithms [3]. Design optimization of sophisticated structural systems involves many objectives, constraints and variables. Therefore, creation of a detailed optimization model incorporating, simultaneously, all the relevant design features is virtually impossible. Researchers and engineers rely on simplified models which provide a fairly accurate approximation of the real structure behaviour. This chapter presents some of the underlying concepts of applying optimization theory for enhancing the stability, dynamic and aeroelastic performance of functionally graded material *(FGM)* structural members. Such concept of *FGM*, in which the properties vary spatially within a structure, was originated in Japan in 1984 during the space project, in the form of proposed thermal barrier material capable of withstanding high temperature gradients. *FGMs* may be defined as advanced composite materials that fabricated to have graded variation of the relative volume fractions of the constituent materials. Commonly, these materials are made from particulate composites where the volume fraction of particles varies in one direction, as shown in Figure 2, or several directions for certain applications. *FGMs* may also be developed using fiber reinforced layers with a volume fractions of fibers changing, rather than constant, producing grading of the material with favorable properties.

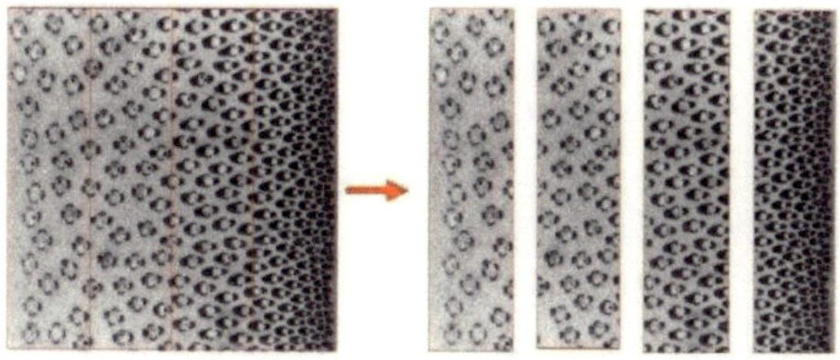

Figure 2. The concept of material grading

Table 1 summarizes the mathematical formulas for determining the equivalent mechanical and physical properties for known type and volume fractions of the fiber and matrix materials [4]. The factor ξ is called the reinforcing efficiency and can be determined experimentally for specified types of fiber and matrix materials. Experimental results fall within a band of $1<\xi<2$. Usually, ξ is taken as 100% for theoretical analysis procedures, especially in case of glass and carbon composites. The 1 and 2 subscripts denote the principal directions of an orthotropic lamina, defined as follows: direction (1): principal fiber direction, also called fiber longitudinal direction; direction (2): In-plane direction perpendicular to fibers, transversal direction.

Property	Mathematical formula*
Young's modulus in direction (1) E_{11}	$E_m V_m + E_{1f} V_f$
Young's modulus in direction (2) E_{22}	$E_m (1+\xi\eta V_f)/(1-\eta V_f)$; $\eta=(E_{2f}-E_m)/(E_{2f}+\xi E_m)$
Shear modulus G_{12}	$G_m (1+\xi\eta V_f)/(1-\eta V_f)$; $\eta=(G_{12f}-G_m)/(G_{12f}+\xi G_m)$
Poisson's ratio ϑ_{12}	$\vartheta_m V_m + \vartheta_{12f} V_f$
Mass density ρ	$\rho_m V_m + \rho_f V_f$
*Subscripts *"m"* and *"f"* refer to properties of matrix and fiber materials, respectively. Assuming no voids are present, then $V_m+V_f=1$.	

Table 1. Halpin-Tsai semi-empirical relations for calculating composite properties [4].

An excellent review paper dealing with the basic knowledge and various aspects on the use of *FGMs* and their wide applications was given in [5]. It was shown that *FGMs* can be promising in several applications such as, spacecraft heat shields, high performance structural elements and critical engine components. A few studies have addressed the dynamics and stability of *FGM* structures. Closed-form expressions for calculating the natural frequencies of an axially graded beam were derived in [6]. The modulus of elasticity was taken as a polynomial of the axial coordinate along the beam's length, and an inverse problem was solved to find the stiffness and mass distributions so that the chosen polynomial serve as an exact mode shape. Another work [7] considered stability of FGM-structures and derived closed form solution for the mode shape and the buckling load of an axially graded cantilevered column. A semi-inverse method was employed to obtain the spatial distribution of the elastic modulus in the axial direction. In reference [8], the buckling of simply supported three-layer circular cylindrical shell under axial compressive load was considered. The middle layer sandwiched with two isotropic layers was made of an isotropic *FGM* whose Young's modulus varies parabolically in the thickness direction. Classical shell theory was implemented under the assumption of very small thickness/radius and very large length/radius ratios. Numerical results showed that the buckling load increases with an increase in the average value of Young's modulus of the middle layer. In the field of structural optimization, reference [9] considered frequency optimization of a cantilevered plate with variable volume fraction according to simple power-laws. Genetic algorithms was implemented to find the optimum values of the power exponents, which maximize the natural frequencies, and concluded that the volume fraction needs to be varied in the longitudinal direction of the plate rather than in the thickness direction. A direct method was proposed in [10] to optimize the natural frequencies of functionally graded beam with variable volume fraction of the constituent materials in the beam's length and height directions. A piecewise bi-cubic interpolation of volume fraction values specified at a finite number of grid points was used, and a genetic algorithm code was applied to find the needed optimum designs. It is the main aim of this chapter to present some fundamental issues concerning design optimization of different types of functionally graded composite structures. Practical realistic optimization models using different strategies for enhancing stability, structural dynamics, and aeroelastic performance are presented and discussed. Design variables represent material type, structure geometry as well as cross sectional parameters. The mathematical formulation is based on dimensionless quantities; therefore the analysis can be valid for different configurations and sizes. Such normalization has led to a naturally scaled optimization models, which is favorable for most optimization techniques. Case studies concerning optimization of FGM composite structures include buckling of flexible columns, stability of thin-walled cylinders subject to external pressure, frequency optimization of FGM bars in axial motion, and critical velocity maximization in pipe flow as a measure of raising the stability boundary. The use of the concept of material grading for enhancing the aeroelastic stability of composite wings have been also addressed. Several design charts that are useful for direct determination of the optimal values of the design variables are introduced. In all, the given mathematical models can be regarded as useful design tools which may save designers from having to choose the values of some of their variables arbitrarily.

2. Buckling optimization of elastic columns

The consideration of buckling stability of elastic columns can be crucial factor in designing efficient structural components. In references [11, 12], optimization models of the strongest columns were developed for maximizing the critical buckling load under equality mass constraint. Emphasizes were given to thin-walled tubular sections, which are more economical than solid sections in resisting compressive loads. The given formulation considered columns made of uniform segments with different material properties, cross-sectional parameters and length, as shown in Figure 3. The simplest problem of equilibrium of a column compressed by an axial force, P, was first formulated and solved by the great mathematician L. Euler in the middle of the eighteenth century. The associated 4th-order governing differential equation in dimensionless form:

$$\hat{w}'''' + \mathrm{P_k}^2\hat{w} = 0, \qquad \mathrm{P_k} = \sqrt{\hat{P}/\hat{E}_k\hat{I}_k}\,,\ k = 1,2,....Ns \tag{2}$$

where ( )′ means differentiation with respect to the dimensionless coordinate $\hat{x}$ and N_s is the total number of segments. The various dimensionless quantities denoted by (^) are defined in Table 2. Equation (2) must be satisfied in the interval $0 \le \bar{x} \le L_k$, where $\bar{x} = \hat{x} - \hat{x}_k$. Its general solution is:

$$\hat{w}(\bar{x}) = a_1 \sin\, P_k\bar{x} + a_2 \cos\, P_k\bar{x} + a_3\bar{x} + a_4 \tag{3}$$

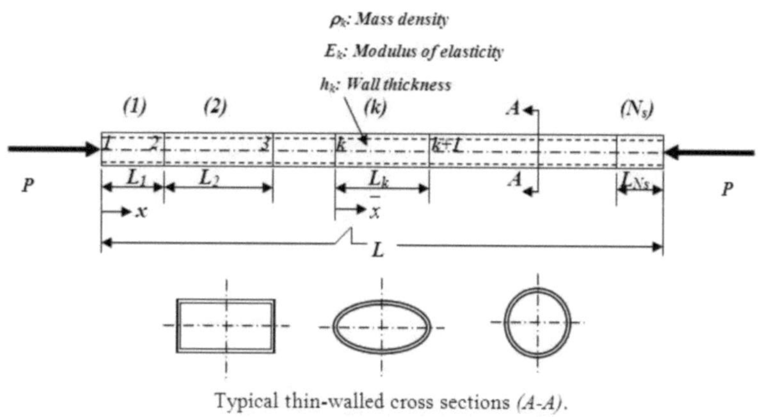

Figure 3. General configuration of a piecewise axially graded thin-walled column

The coefficients a_i's in Equation (3) can be expressed in terms of the state variables at both nodes of the *Kth* segment, which results in the following matrix relation:

$$\begin{Bmatrix} \hat{w}_{k+1} \\ \varphi_{k+1} \\ \hat{M}_{k+1} \\ \hat{F}_{k+1} \end{Bmatrix} = \begin{bmatrix} 1 & \frac{-S_k}{P_k} & \frac{-(1-C_k)}{\hat{P}} & \left(\frac{S_k}{\hat{P}P_k} - \frac{\hat{L}_k}{\hat{P}}\right) \\ 0 & C_k & \frac{P_k S_k}{\hat{P}} & \frac{(1-C_k)}{\hat{P}} \\ 0 & \frac{-\hat{P}S_k}{P_k} & C_k & \frac{S_k}{P_k} \\ 0 & 0 & 0 & 1 \end{bmatrix} \begin{Bmatrix} \hat{w}_k \\ \phi_k \\ \hat{M}_k \\ \hat{F}_k \end{Bmatrix} \tag{4}$$

where $S_k = Sin P_k L_k$ and $C_k = Cos P_k L_k$. Applying Equation (4) successively to all the segments composing the column and taking the products of all the resulting matrices, the state variables at both ends of the column can be related to each other through an overall transfer matrix. Therefore, by the application of the appropriate boundary conditions and consideration of the non-trivial solution, the associated characteristic equation for determining the critical buckling load can be accurately obtained. The exact buckling analysis outlined above can be coupled with a standard nonlinear mathematical programming algorithm for the search of columns designs with the largest possible resistance against buckling. It is important to bear in mind that design optimization is only as meaningful as its core structural analysis model. Any deficiencies therein will certainly be reflected in the optimization process.

Quantity	Non-dimensionalization*
Axial coordinate	$\hat{x} = x / L$
Length of *Kth* segment	$\hat{L}_k = L_k / L$
Transverse deflection	$\hat{w} = w / L$
Wall thickness	$\hat{h}_k = h_k / h$
Second moment of area	$\hat{I}_k = I_k / I \ (= \hat{h}_k)$
Modulus of elasticity	$\hat{E}_k = E_k / E$
Bending moment	$\hat{M} = M * (L / EI)$
Shearing force	$\hat{F} = F * (L^2 / EI)$
Axial force	$\hat{P} = P * (L^2 / EI)$
Mass density	$\hat{\rho}_k = \rho_k / \rho$
Total structural mass	$\hat{M}_s = \sum_{k=1}^{Ns} \hat{\rho}_k \hat{h}_k \hat{L}_k$
*Baseline design parameters: *L*=total column's length, *h*=wall thickness, *I*= second moment of area, *E*=modulus of elasticity = $(E_f + E_m)/2$, ρ=mass density= $(\rho_f + \rho_m)/2$.	

Table 2. Definition of dimensionless quantities

Therefore, the strongest column design problem may be cast in the following:

Maximize

$$\hat{P}_{cr}$$

Subject to

$$\hat{M}_s = 1$$

$$\sum_{k=1}^{Ns} \hat{L}_k = 1 \tag{5}$$

Side constraints are always present by imposing lower and upper limits on the design variables to avoid having odd-shaped unrealistic column design in the final optimum solutions. Reference [12] presented optimum patterns for cases of simply supported and cantilevered FGM columns constructed from unidirectional fibrous composites with properties given in Table 3. The case of a symmetrical simply supported (*S.S.*) column made of E-glass/epoxy and constructed from different number of segments is depicted in Figure 4. For the case of 6-segments, the optimum zone of the dimensionless critical buckling load augmented with the mass equality constraint was determined and found to be well behaved in the selected $(V_A - L)_3$ design space. Referring to Figure 5, three distinct regions can be observed: two empty regions to the left and right violating the mass equality constraint and the middle feasible region containing the global optimum solution. It is also seen that the optimal feasible domain is bounded from left and right by two heavy zigzagged lines owing to the fact that many contours are stuck to these borderlines and are not allowed to penetrate them for not violating the imposed mass constraint. The final optimum design point was found to be $(V_A, \hat{L})_{k=1,2,3} = (0.095, 0.0875), (0.38125, 0.140625), (0.69177, 0.271875)$ where $(P_{cr})_{max} = 11.649375$. This represents an optimization gain of about *18.0%* relative to the baseline value (π^2). Other types of materials were also addressed in [12] including, carbon/epoxy, S-glass/epoxy, E-glass/Vinyl-ester and S-glass/Vinyl-ester. In all cases the

Composite material	material *(A)* = Fibers		material *(B)* = matrix	
	$\rho_A (g/cm^3)$	E_A *(GPa)*	$\rho_B (g/cm^3)$	$E_B (GPa)$
E-glass/epoxy	2.54	73.0	1.27	4.3
S-glass/epoxy	2.49	86.0		
Carbon/epoxy	1.81	235.0		
E-glass/Vinylester	2.54	73.0	1.15	3.5
S-glass/Vinylester	2.49	86.0		

Table 3. Material properties of selected fiber-reinforced composites [12]

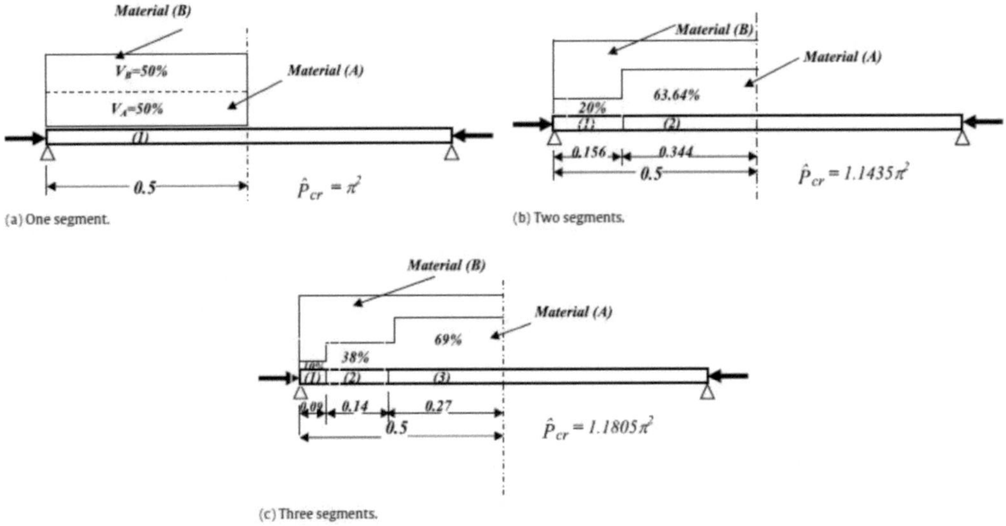

Figure 4. Optimum simply supported columns with piecewise axial material grading

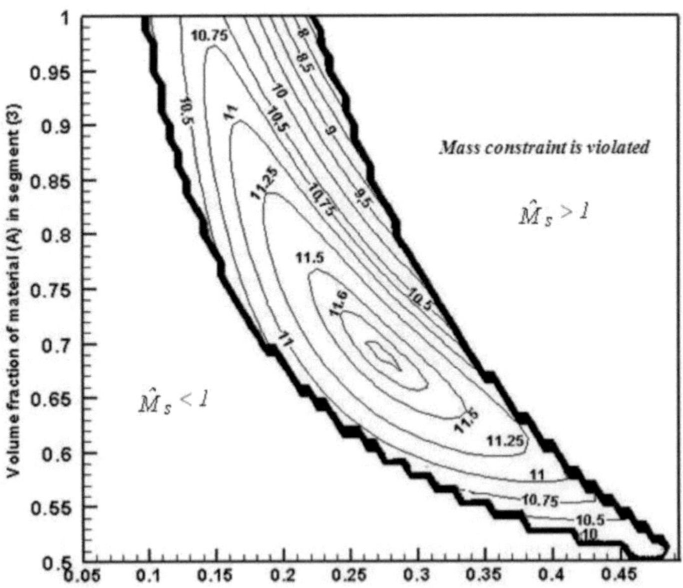

Figure 5. Optimum zone for a symmetrical 6-segment *S.S.* columns made of E-glass/epoxy composites.

buckling load was found to be very sensitive to variation in the segment length. Investigators who use approximate methods, such as finite elements, have not recognized that the length of each element can be taken as a main optimization variable in addition to the cross-sectional properties. The increase in the number of segments would, naturally, result in higher values of the dimensionless critical buckling load. However, care ought to be taken for the corresponding increase in cost due to the resulting complications in the associated assembling and manufacturing procedures.

Other cases of cantilevered columns were also investigated. The associated boundary conditions are: at $\hat{x}=0$ $\hat{w}=\varphi=0$, and at $\hat{x}=1.0$ $\hat{M}=\hat{F}=0$. Figure 6 shows the attained optimal solutions for cantilevered columns made of unidirectional E-glass/epoxy composites and constructed from different number of segments (N_s). For a three-segment column, the global optimal solution was found to be $(P_{cr})_{max}=2.90938$ occurring at the design point $(V_A,L_k)_{k=1,2,3}$ = (0.70, 0.514), (0.4125, 0.2785), (0.122,0.2075). This means that the strongest column made of only three segments can withstand a buckling load 18% higher than that with uniform mass and stiffness distributions, which represents a truly optimized column design. In fact, the exact buckling load can be obtained for any number of segments, type of cross section and type of boundary conditions. The given multi-segment model has the advantageous of achieving global optimality for the strongest columns shape that can be manufactured economically from any arbitrary number of segments. Sensitivity of the design variables on the buckling load should be included in a more general formulation.

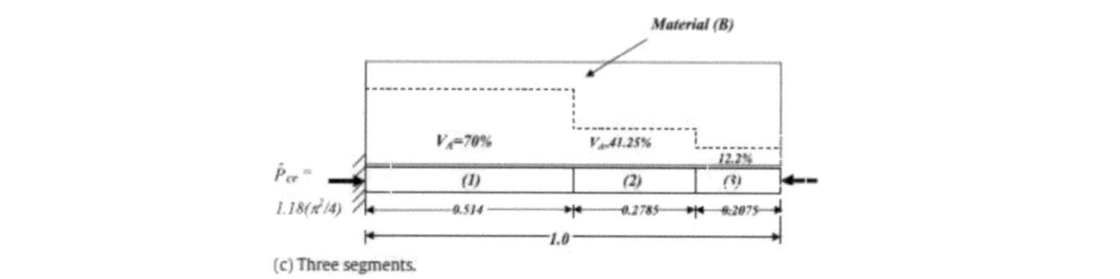

Figure 6. Strongest cantilevered columns with axial material grading: Material (A)=E-glass fibers, material (B)=epoxy matrix

3. Stability of FGM long cylinders under external pressure

A common application of composites is the design of cylindrical shells under the action of external hydrostatic pressure, which might cause collapse by buckling instability. Examples are the underground and underwater pipelines, rocket motor casing, boiler tubes subjected to external steam pressure, and reinforced submarine structures. The composite cylindrical vessels for underwater applications are intended to operate at high external hydrostatic pressure (sometimes up to 60 MPa). For deep- submersible long-unstiffened vessels, the hulls are generally realized using multilayered, cross-ply, composite cylinders obtained following the filament winding process. Previous numerical and experimental studies have shown that failure due to structural buckling is a major risk factor for thin laminated cylindrical shells. Figure 7 shows the structural model used in reference [13], where the effect of changing the fiber volume fraction in each lamina was taken in the formulation of the structural model.

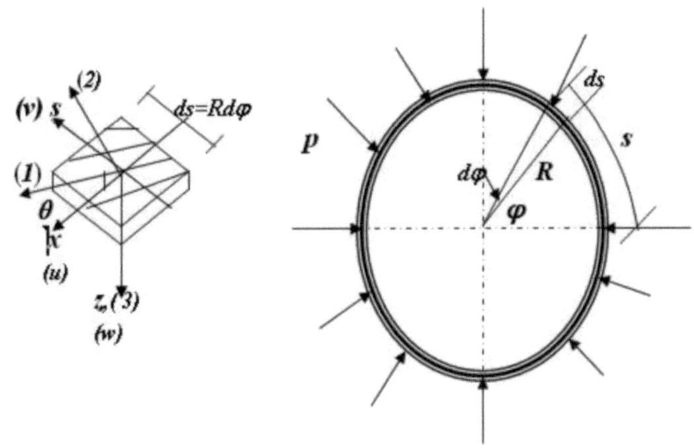

Figure 7. Laminated composite shell under external pressure (*u* displacement in the axial direction *x*, *v* in the tangential direction *s*, *w* in the radial direction *z*)

The governing differential equations of anisotropic rings/long cylinders subjected to external pressure are cast in the following [13]:

$$M'_{ss} + R(N'_{ss} - \beta N_{ss}) = \beta\, pR^2$$

$$M''_{ss} - R[N_{ss} + (\beta N_{ss})' + p(w_o + v'_o)] = pR^2 \qquad (6)$$

where the prime denotes differentiation with respect to angular position φ, and $\beta = (v_o - w'_o)/R$. Two possible solutions for Eq. (6) can be obtained; one for the pre-buckled state and the other termed as the bifurcation solution obtained by perturbing the displacements about the pre-buckling solution. For laminated composite rings and long cylindrical shells the only significant strain components are the hoop strain (ε^0_{ss}) and the circumferential curvature (κ_{ss}) of the mid-surface. In the case of thin rings the axial and shear forces (N_{xx}, N_{xs}) must vanish along the free edges. The bending and twisting moments (M_{xx}, M_{xs}) may also be neglected. The final closed form solution for the critical buckling pressure is given by the following mathematical expression:

$$p_{cr} = 3\left[\frac{D_{ani}}{R^3}\right]\left[\frac{1-(\psi^2/\alpha)}{1+\alpha+2\psi}\right], \quad \psi = (\frac{1}{R})(\frac{B_{ani}}{A_{ani}}), \quad \alpha = (\frac{1}{R^2})(\frac{D_{ani}}{A_{ani}}) \tag{7}$$

which is only valid for thin rings/cylinders with thickness-to-radius ratio $(h/R) \leq 0.1$. The stiffness coefficients A_{ani}, B_{ani} and D_{ani} are calculated, for the case of long cylinders from:

$$\begin{bmatrix} A_{ani} & B_{ani} \\ B_{ani} & D_{ani} \end{bmatrix}_{cylinder} = \begin{bmatrix} A_{22} & B_{22} \\ B_{22} & D_{22} \end{bmatrix} \tag{8}$$

And for circular rings:

$$\begin{bmatrix} A_{ani} & B_{ani} \\ B_{ani} & D_{ani} \end{bmatrix}_{ring} = \begin{bmatrix} A_{22} & B_{22} \\ B_{22} & D_{22} \end{bmatrix} - [S_2]^T [S_1]^{-1} [S_2] \tag{9}$$

where
$$[S_1] = \begin{bmatrix} A_{11} & A_{16} & B_{11} & B_{16} \\ A_{16} & A_{66} & B_{16} & B_{66} \\ B_{11} & B_{16} & D_{11} & D_{16} \\ B_{16} & B_{66} & D_{16} & D_{66} \end{bmatrix} \quad \text{and} \quad [S_2] = \begin{bmatrix} A_{12} & B_{12} \\ A_{26} & B_{26} \\ B_{12} & D_{12} \\ B_{26} & D_{26} \end{bmatrix} \tag{10}$$

A_{ij} are called the extensional stiffnesses given by:

$$A_{ij} = h\sum_{k=1}^{n} (\overline{Q}_{ij})_k (\hat{z}_k - \hat{z}_{k-1}) \tag{11}$$

B_{ij} are called the bending-extensional stiffnesses given by:

$$B_{ij} = \frac{h^2}{2}\sum_{k=1}^{n} (\overline{Q}_{ij})_k (\hat{z}_k^2 - \hat{z}_{k-1}^2) \tag{12}$$

D_{ij} are called the bending stiffnesses:

$$D_{ij} = \frac{h^3}{3}\sum_{k=1}^{n} (\overline{Q}_{ij})_k (\hat{z}_k^3 - \hat{z}_{k-1}^3) \tag{13}$$

where $\hat{z}_k = z_k / h$ is a dimensionless coordinate, and $\hat{h}_k = \hat{z}_k - \hat{z}_{k-1}$ is the dimensionless thickness of the *kth* lamina. The associated optimization problem shall seek maximization of the critical buckling pressure p_{cr} while maintaining the total structural mass constant at a value equals to that of a reference baseline design. Optimization variables include the fiber volume fraction (V_{fk}), thickness (h_k) and fiber orientation angle (θ_k) of the individual *k-th* ply,

$k=1, 2, \ldots n$ (total number of plies). Side constraints are always imposed on the design variables for geometrical, manufacturing or logical reasons to avoid having unrealistic odd shaped optimum designs. The first case study to be examined herein is a long thin-walled cylindrical shell fabricated from E-glass/epoxy composites with the lay-up made of only two plies $(n=2)$ having fibers parallel to the x-axis (i.e. $\theta_1=\theta_2=0$). Considering the case with no side inequality constraints imposed on the design variables, Figure 8 shows the developed $\hat{P}_{cr}$ -level curves, augmented with the mass equality constraint, in $(V_{f1}-\hat{h}_1)$ design space.

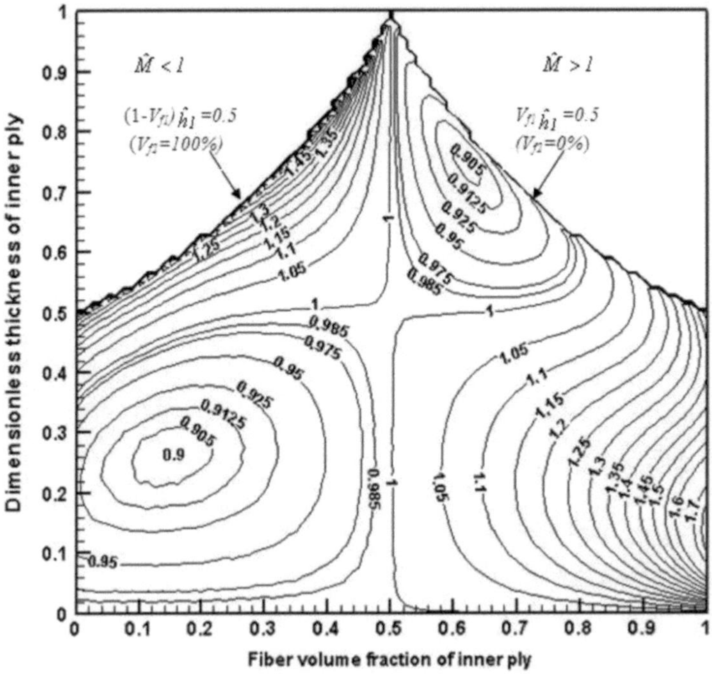

Figure 8. Optimum design space containing p_{cr}–isobars augmented with the mass equality constraint $\hat{M}=1.0$. Case of two-layer, E-glass/epoxy cylinder with fibers parallel to cylinder axis ($\theta_1=\theta_2=0$).

It is seen that such a constrained objective function is well behaved in the selected design space having the shape of a tent with its ceiling formed by two curved lines, above which the mass equality constraint is violated. Their zigzagged pattern is due to the obliged turning of many contours, which are not allowed to penetrate the tent's ceiling and violate the mass equality constraint. The curve to the left represent a 100% fiber volume fraction of the outer ply, V_{f2}, while the other curve to the right represents zero volume fraction, that is $V_{f2}=0\%$. Two local minima with p_{cr} near a value of 0.90 can be observed: one to the lower left zone near the design point $(V_{fk},h_k)_{k=1,2}$ = (0.15, 0.25), (0.6165, 0.75) while the other lies at the upper right zone close to the point (0.625, 0.745), (0.135, 0.255). This represents degradation in the stability level by about 10.6% below the baseline value. On the other hand, the unconstrained absolute optimum value of the dimensionless critical buckling pressure was found to be 1.7874 at the design point (1.0, 0.145), (0.415, 0.855). A more realistic optimum design has been obtained by imposing the side constraints: $0.25 \leq V_{fk} \leq 0.75$, $k=1, 2$. The attained solution is $(p_{cr})_{max}$ =1.2105 at the design point (0.75, 0.215), (0.4315, 0.785), showing that good shell designs with higher stability level ought to have a thinner inner layer with higher fiber volume fraction and a thicker outer layer with less volume fraction. To see the effect of the ply angle, another case of study has been considered for a cylinder constructed from two balanced plies $(\pm\theta)$ with equal thicknesses and same material properties of E-glass/epoxy composites. This type of stacking sequence is widely used in filament wound circular shells since such a manufacturing process inherently dictates adjacent $(\pm\theta)$ layers. A local minimum was found near the design point (V_{f1},θ) = $(0.375, 0.0)$ with p_{cr}=0.9985, indicating a degradation in the stability level below the baseline design. The absolute maximum occurred at the design points (V_{f1},θ)=$(0.5,\pm 90^o)$ *with* $(p_{cr})_{max}$=3.45766, which means that the dimensional critical pressure, p_{cr}=3.45766x2.865=9.906 $x(h/R)^3$ *GPa*. Figure 9 depicts the final global optimum designs of cylinders constructed from adjacent $(+\theta)$ and $(-\theta)$ plies for the different types of the selected composite materials. All shall have the same optimal solution $(V_{fk},h_k)_{k=1,2}$ = (0.75, 0.215), (0.4315, 0.785), independent upon the shell thickness-to radius ratio (h/R), a major contribution of the given formulation.

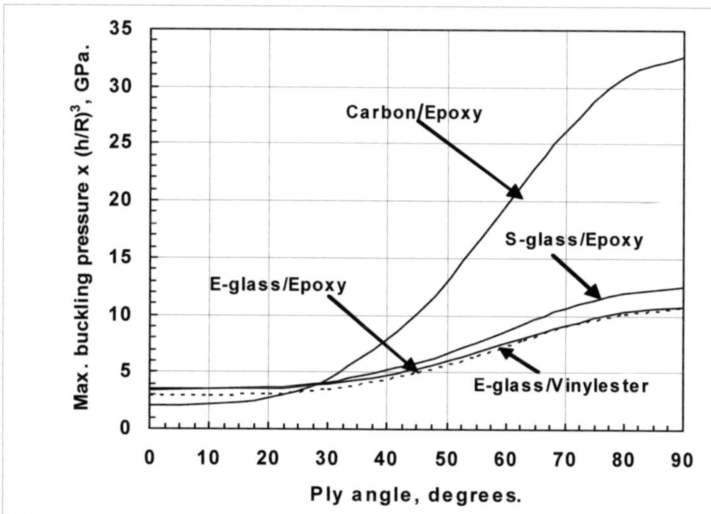

Figure 9. Variation of the absolute maximum buckling pressure with ply angle for balanced $(\pm\theta^o)$ cylinders with structural mass preserved constant

Other cases of study include optimization of two different constructions of multi-layered cylinders made of *AS-4* carbon/epoxy composites. The first one is called a lumped-layup construction with the inner half of its wall composed of 90° hoop layers and the outer half made of *±20°* helically wound layers. The second type has different stacking sequence where the *±20°* layers are sandwiched in between outer and inner *90°* hoop layers. Optimum solutions are given in Table 4, indicating that good designs shall have thicker hoop wound layers with higher volume fraction of the fibers near the upper limiting values imposed by the manufacturers. On the other hand, the sandwiched helically wound layers are seen to be thinner and have less fiber volume fractions.

<table>
<tr><th rowspan="3">(h/R)</th><th colspan="2">[90⁰/±20⁰] layup</th><th colspan="2">[90⁰/±20⁰/90⁰] layup</th></tr>
<tr><th colspan="2">$p_{cr,max}$=9.37x(10h/R)^3 MPa</th><th colspan="2">$p_{cr,max}$=36.634x(10h/R)^3 MPa</th></tr>
<tr><th>Optimum value</th><th>% gain</th><th>Optimum value</th><th>% gain</th></tr>
<tr><td>1/50</td><td>0.075</td><td>17.19</td><td>0.293</td><td>26.84</td></tr>
<tr><td>1/25</td><td>0.596</td><td>15.5</td><td>2.344</td><td>26.84</td></tr>
<tr><td>1/20</td><td>1.171</td><td>15.6</td><td>4.579</td><td>26.91</td></tr>
<tr><td>1/15</td><td>2.776</td><td>14.81</td><td>10.854</td><td>26.92</td></tr>
<tr><td colspan="2">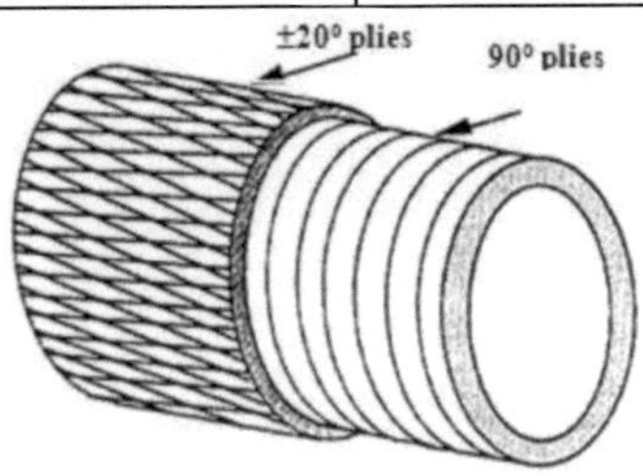
</td><td colspan="3">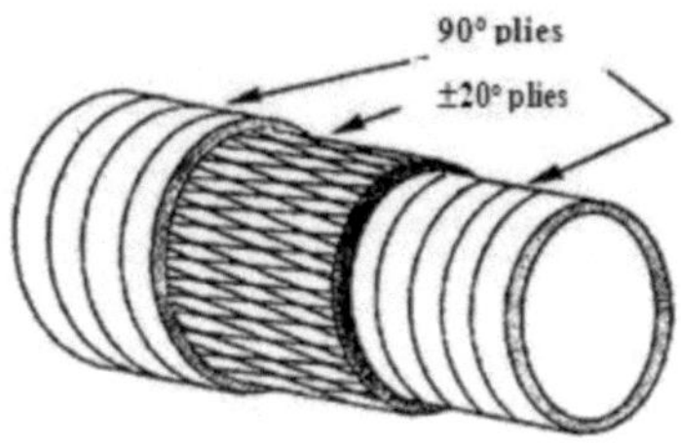
</td></tr>
<tr><td>Optimum solutions
Two helical layers
(V_f, h, θ)=</td><td colspan="2">(0.25, 0.225, ±20⁰)</td><td colspan="2">(0.2925, 0.235, ±20⁰)</td></tr>
<tr><td>Two hoop layers
(V_f, h, θ)=</td><td colspan="2">(0.705, 0.275, 90⁰)</td><td colspan="2">(0.6835, 0.265, 90⁰)</td></tr>
</table>

Table 4. Optimum buckling design of multi-layered, AS-4, FGM composite cylinders

4. Dynamic optimization of FGM bars in axial motion

Elastic slender bars in axial motion can give rise to significant vibration problems, which assesses the importance of considering optimization of natural frequencies. These frequencies, besides being maximized, must be kept out of the range of the excitation frequencies in order to avoid large induced stresses that can exceed the reserved fatigue strength of the materials and, consequently, cause failure in a short time. Expressed mathematically, two different design criteria are implemented here for optimizing frequencies:

Frequency-placement criterion: Minimize $\sum_i W_{fi}\omega_i$ (14)

Maximum-frequency criterion: Maximize $\sum_i W_{fi}\omega_i$ (15)

In both criteria, an equality constraint should be imposed on the total structural mass in order not to violate other economic and performance requirements. Equation (10) represents a weighted sum of the squares of the differences between each important frequency ω_i and its desired (target) frequency ω_i^*. Appropriate values of the target frequencies are usually chosen to be within close ranges (called frequency windows) of those corresponding to a reference or baseline design, which are adjusted to be far away from the critical exciting frequencies. The main idea is to tailor the mass and stiffness distributions in such a way to make the objective function a minimum under the imposed mass constraint. The second alternative for reducing vibration is the direct maximization of the system natural frequencies as expressed by equation (11). Maximization of the natural frequencies can ensure a simultaneous balanced improvement in both of stiffness and mass of the vibrating structure. It is a much better design criterion than minimization of the mass alone or maximization of the stiffness alone. The latter can result in optimum solutions that are strongly dependent on the limits imposed on either the upper values of the allowable deflections or the acceptable values of the total structural mass, which are rather arbitrarily chosen. The proper determination of the weighting factors W_{fi} should be based on the fact that each frequency ought to be maximized from its initial value corresponding to a baseline design having uniform mass and stiffness properties. Reference [14] applied the concept of material grading for enhancing the dynamic performance of bars in axial motion. The associated eigenvalue problem is cast in the following:

$$\hat{E}\frac{d^2\hat{U}}{d\hat{x}^2}+\frac{d\hat{E}}{d\hat{x}}\cdot\frac{d\hat{U}}{d\hat{x}}+\hat{\rho}\hat{\omega}^2\hat{U}=0,\ \ 0<\hat{x}<1 \tag{16}$$

where $\hat{U}=U/L$ is the dimensionless amplitude and $\hat{\omega}=\omega L\sqrt{\rho/E}$ dimensionless frequency. Both continuous and discrete distributions of the volume fractions of the selected composite material were analyzed in [14]. The general solution of Eq. (12), where the modulus of elasticity and mass density vary in the axial direction, can be expressed by the following power series:

$$\hat{U}(\hat{x})=\sum_{m=1}^{2}C_m\lambda_m(\hat{x}) \tag{17}$$

where C_m's are the constants of integration and λ_m's are two linearly independent solutions that have the form:

$$\lambda_m(\hat{x})=\sum_{n=m}^{\infty}a_{m,n}\hat{x}^{n-1}\quad (n\geq m) \tag{18}$$

The unknown coefficients $a_{m,n}$ can be determined by substitution into the differential equation (12) and equating coefficients of like powers of $\hat{x}$. Table 5 summarizes the appropriate mathematical expressions of the frequency equation for any desired case, which can be obtained by application of the associated boundary conditions and consideration of nontrivial solutions.

Variation of the volume fractions in *FGM* structures is usually described by power-law distributions. Figure 10 shows both linear and parabolic models for material grading along the bar span. Results given in [14] showed that, for Fixed-Fixed and Fixed-Free boundary conditions, patterns with higher fiber volume fraction near the fixed ends are always

	Boundary conditions	Frequency equation	$(\hat{\omega}_o)_i$
Fixed-Fixed Bar Symmetrical modes	$\hat{U}(0)=\hat{U}'(1/2)=0$	$\sum_{n=3}^{\infty} a_{2n}(n-1)/(2)^{n-2}=-1$	$(\pi, 3\pi, 5\pi)$
Unsymmetrical modes	$\hat{U}(0)=\hat{U}(1/2)=0$	$\sum_{n=2}^{\infty} a_{2n}/(2)^{n-1}=0$	$(2\pi, 4\pi, 6\pi)$
Fixed-Free Bar	$\hat{U}(0)=\hat{U}'(1)=0$	$\sum_{n=3}^{\infty} a_{2n}(n-1)=-1$	$(\pi, 3\pi, 5\pi)/2$
Free-Free Bar Symmetrical modes	$\hat{U}'(0)=\hat{U}'(1/2)=0$	$\sum_{n=3}^{\infty} a_{1n}(n-1)/(2)^{n-2}=0$	$(2\pi, 4\pi, 6\pi)$
Unsymmetrical modes	$\hat{U}'(0)=\hat{U}(1/2)=0$	$\sum_{n=3}^{\infty} a_{1n}/(2)^{n-1}=-1$	$(\pi, 3\pi, 5\pi)$

Table 5. Frequency equations for different types of boundary conditions. $\hat{\omega}_{o,i}$ are the dimensionless natural frequencies of the baseline design ($\hat{\omega}_o$=0 corresponding to the first rigid body mode of a Free-Free bar). The notation $(\ )'$ means $d/d\hat{x}$.

favorable. The opposite trend is true for cases of Free-Free bars. Maximization of the fundamental frequency alone produces an optimization gain of about 14.33% for the linear model with 0% and 100% volume fractions at the ends of the optimized bars with different boundary conditions. However, a drastic reduction in the 2nd and 3rd frequencies was observed. Better solutions have been achieved by maximizing a weighted-sum of the first three frequencies, where the parabolic model was found to excel the linear one in producing balanced improvements in all frequencies. Results have also indicated that the Fixed-Fixed bars are recommended to have concave distribution rather than convex one. The latter produce poor patterns with degraded stiffness-to-mass ratio levels. The opposite trend was observed for the free-free bars, where the convex type is much more favorable than the concave type. Both concave and convex shapes can be accepted for a cantilevered bar. For piecewise models, the developed isomerits for the case of Free-Free bar built of four symmetrical segments made of carbon/epoxy composites are shown in Figure 11. The global maximum of the fundamental frequency is located at the lower region to the left of the design space having a value of $\hat{\omega}_{1,max}=3.45406$ at the optimal design point $(V_A, \hat{L})_{k=1,2}$ =(0.1885,0.1625), (0.650, 0.3375), which represents about 10% optimization gain.

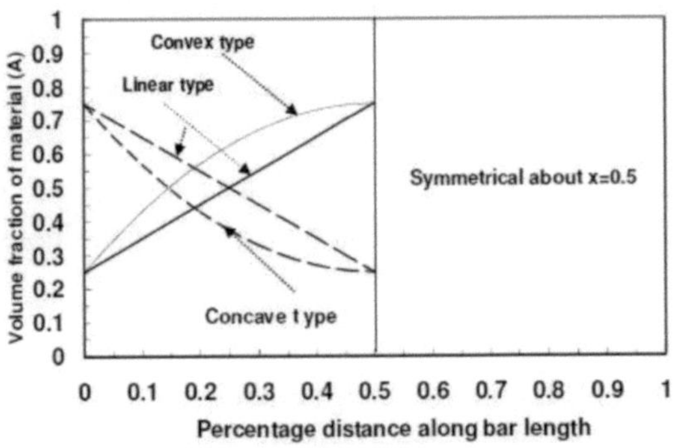

Figure 10. Symmetrical shape models of volume fraction distribution along bar length

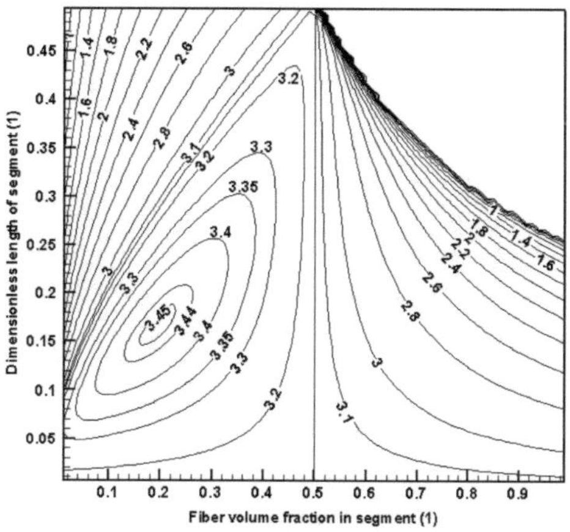

(a) Fundamental frequency (Unsymmetrical mode)

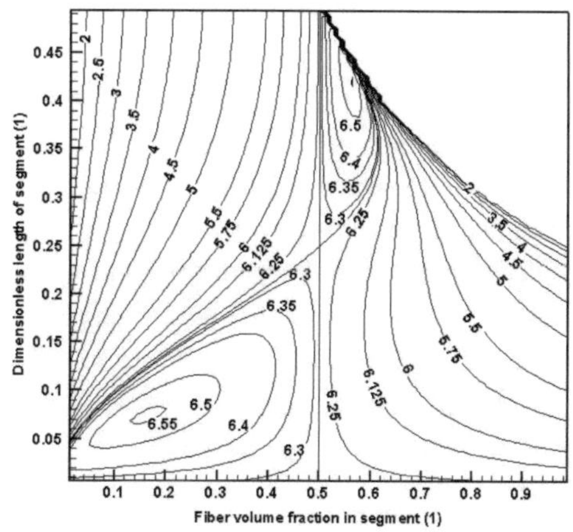

(b) Second frequency (Symmetrical mode).

Figure 11. Dimensionless frequency isomerits of free-free bar under mass constraint

5. Material grading for improved aeroelastic stability of composite wings

Aircraft wings can experience aeroelastic instability condition in high speed flight regimes. A solution that can be promising to enhance aeroelastic stability of composite wings is the use of the concept of functionally graded materials (FGMs) with spatially varying properties. Reference [15] introduced some of the underlying concepts of using material grading in optimizing subsonic wings against torsional instability. Exact mathematical approach allowing the material properties to change in the wing spanwise direction was applied, where both continuous and piecewise structural models were successfully implemented. The enhancement of the torsional stability was measured by maximization of the critical flight speed at which divergence occurs with the total structural mass kept at a constant value in order not to violate other performance requirements. Figure 12 shows a rectangular composite wing model constructed from uniform piecewise panels, where the design variables are defined to be the fiber volume fraction (V_f) and length (L) of each panel.

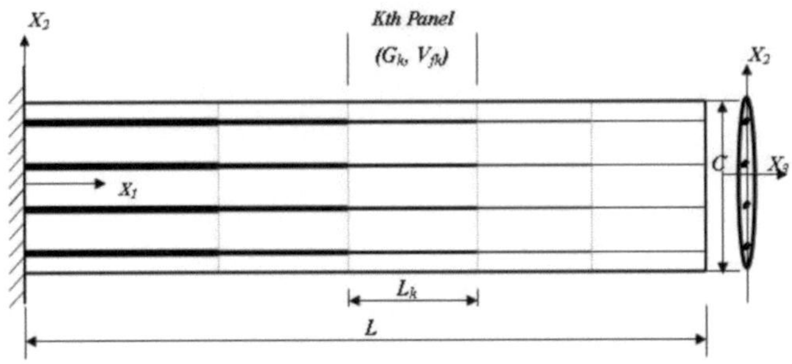

Figure 12. Composite wing model with material grading in spanwise direction

The isodiverts (lines of constant divergence speed) for a wing composed from two panels made of carbon-AS4/epoxy-3501-6 composite are shown in Figure 13. The selected design variables are (V_{f1},L_1) and (V_{f2},L_2). However, one of the panel lengths can be eliminated, because of the equality constraint imposed on the wing span. Another variable can also be discarded by applying the mass equality constraint, which further reduces the number of variables to only any two of the whole set of variables. Actually the depicted level curves represent the dimensionless critical flight speed augmented with the imposed equality mass constraint. It is seen that the function is well behaved, except in the empty regions of the first and third quadrants, where the equality mass constraint is violated. The final constrained optima was found to be (V_{f1},L_1)= (0.75, 0.5) and (V_{f2},L_2)= (0.25, 0.5), which corresponds to the maximum critical speed of 1.81, representing an optimization gain of about 15% above the reference value π/2. The functional behavior of the critical flight speed $\hat{V}_{div}$ of a three-panel model is shown in Figure 14, indicating conspicuous design trends for configurations with improved aeroelastic performance. As seen, the developed isodiverts have a pyramidal shape with its vertex at the design point (V_{f2}, L_2)=(0.5, 1.0) having $\hat{V}_{div}$ =π/2. The feasible domain is bounded from above by the two lines representing cases of two-panel wing, with V_{f1}=0.75 for the line to the left and V_{f3}=0.25 for the right line. The contours near these two lines are asymptotical to them in order not to violate the mass equality constraint. The final global optimal solution, lying in the bottom of the pyramid, was calculated using the MATLAB optimization toolbox routines as follows: $(V_{fk}, L_k)_{k=1,2,3}$ = (0.75, 0.43125), (0.5, 0.1375), (0.25, 0.43125) with $\hat{V}_{div}$=1.82, which represents an optimization gain of about 16%. Actually, the given exact mathematical approach ensured the attainment of global optimality of the proposed optimization model. A more general case would include material grading in both spanwise and airfoil thickness directions.

6. Optimization of FGM pipes conveying fluid

The determination of the critical flow velocity at which static or dynamic instability can be encountered is an important consideration in the design of slender pipelines containing

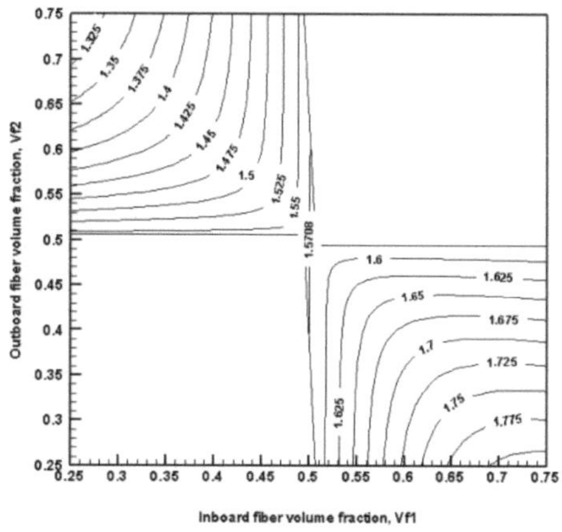

Figure 13. Isodiverts of for a two-panel wing model

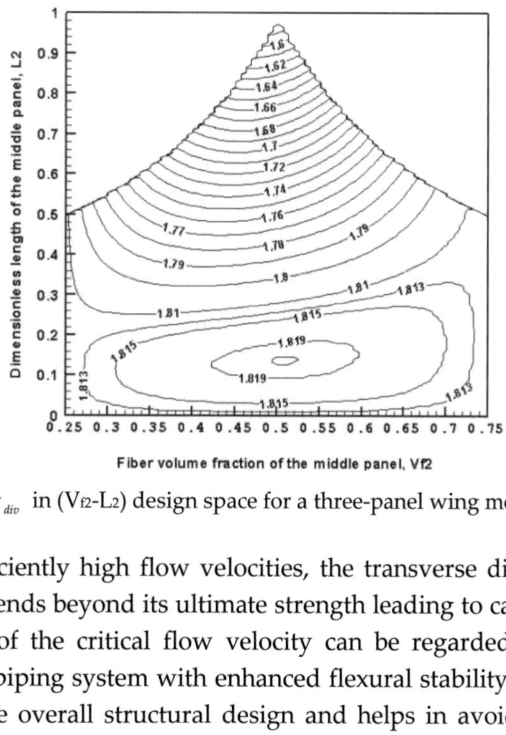

Figure 14. Isodiverts of $\hat{V}_{div}$ in (V_{f2}-L_2) design space for a three-panel wing model

flowing fluid. At sufficiently high flow velocities, the transverse displacement can be too high so that the pipe bends beyond its ultimate strength leading to catastrophic instabilities. In fact maximization of the critical flow velocity can be regarded as a major aspect in designing an efficient piping system with enhanced flexural stability. It can also have other desirable effects on the overall structural design and helps in avoiding the occurrence of large displacements, distortions and excessive vibrations, and may also reduce fretting among structural parts, which is a major cause of fatigue failure. The dynamic characteristics of fluid-conveying functionally graded materials cylindrical shells were investigated in [16]. A power-law was implemented to model the grading of material properties across the shell thickness and the analysis was performed using modal superposition and Newmark's direct time integration method. Reference [17] presented an analytical approach for maximizing the critical flow velocity, also known as divergence velocity, through multi-module pipelines for a specified total mass. Optimum solutions were given for simply supported pipes with the design variables taken to be the wall thickness and length of each module composing the pipeline. A recent work [18] considered stability optimization of FGM pipelines conveying fluid, where a general multimodal model was formulated and applied to cases with different boundary conditions. A more spacious optimization model was given and extending the analysis to cover both effects of material, thickness grading and type of support boundary conditions. The model incorporated the effect of changing the volume fractions of the constituent materials for maximizing the critical flow velocity while maintaining the total mass at a constant value. Additional constraints were added to the optimization model by imposing upper limits on the fundamental eigenvalue to overcome the produced multiplicity near the optimum solution. Figure 15 shows the pipe model under consideration consisting of rigidly connected thin-walled tubes, each of which has different material properties, cross-sectional dimensions and length. The tube thickness, *h,* is assumed to be very small as compared with the mean diameter, *D.* The pipe conveys an incompressible fluid flowing steadily with an axial velocity U_k through the kth module. The variation in the velocity across the cross section was neglected, and the pipe was assumed to be long and slender so that the classical engineering theory of bending can be applicable. The effects of structural damping, damping of surroundings and gravity were not considered. Practical designs ignoring small damping, which has stabilizing effect on the system motion, are always conservatives. The model axis in its un-deformed state coincides with the horizontal x-axis, and the free small motion of the pipe takes place in a two dimensional plane with transverse displacement, *w.*

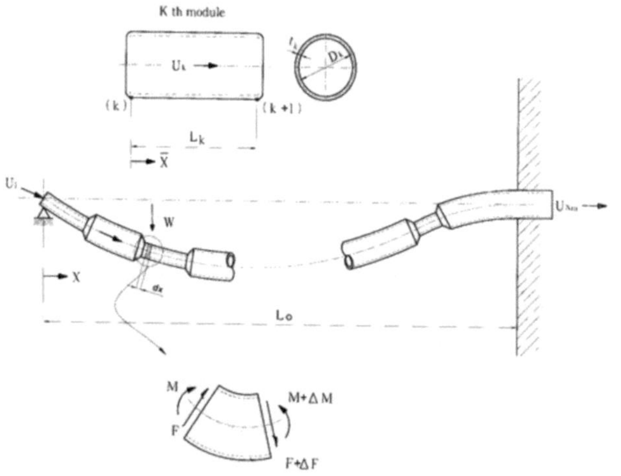

Figure 15. General configuration of a piecewise axially graded pipe conveying fluid.

The various parameters are normalized by their corresponding values of a baseline pipe having the same total mass and length, material and fluid properties, and boundary conditions as well. The baseline pipe has uniform mass and stiffness distributions along its length and is made of two different materials denoted by *(A)* and *(B)* with equal volume fractions *(V),* i.e. V_A=V_B= *50%*. The governing differential equation in dimensionless form:

$$w'''' + \lambda_k^2 w'' = 0 \tag{19}$$

where $\lambda_k = U_k\sqrt{\dfrac{A_k}{E_k I_k}} = \dfrac{U A_{max}}{\sqrt{A_k E_k I_k}}$

$$k=1,2,\ldots,N_m \tag{20}$$

which is valid over the length of any kth module of the pipe, i.e. $0 \le \bar{x} \le L_k$, where $\bar{x} = x - x_k$

In equation (16), U stands for the flow velocity through the pipe module having the maximum cross sectional area A_{max} and N_m is the total number of modules composing the pipeline. It is noted that consideration of the continuity equation provides that $U_k A_k = U A_{max}$ $,k=1,2,\ldots N_m$ Possible boundary conditions at the end supports of the pipeline are stated in the following:

(a) Hinged-Hinged (H/H): $w(0)=w''(0)=0$
$w(1)=w''(1)=0$
(b) Clamped-Hinged (C/H): $w(0)=w''(0)=0$
$w(1)=w'(1)=0$
(c) Clamped-Clamped (C/C): $w(0)=w'(0)=0$
$w(1)=w'(1)=0$

For a cantilevered pipeline, static instability caused by divergence is unlikely to happen. The non-trivial solution of the associated characteristic equation results in a vanishing bending displacement over the entire span of the pipeline. For such pipe configuration, dynamic instability (flutter) may only be considered. The state variable are defined by the vector

$$\underline{Z}_k^T = \left[w\,\varphi\, M\, F \right]_k = \left[w\, {-w'}\, -EIw''\, -EIw''' \right]_k \tag{21}$$

At two successive joints (k) and $(k+1)$ the state vectors are related to each other by the matrix equation

$$\underline{Z}_{k+1} = [\, T_k \,]\, \underline{Z}_k \tag{22}$$

where $[T_k]$ is a square matrix of order 4x4 known as the transmission or transfer matrix of the kth pipe module. For a pipeline built from N_m - uniform modules, Eq.(18) can be applied at successive joints to obtain

$$\underline{Z}_{Nm+1} = [\, T \,]\ \underline{Z}_1 \tag{23}$$

where $[T]$ is called the overall transmission matrix formed by taking the products of all the intermediate matrices of the individual modules. Therefore, applying the boundary conditions and considering only the non-trivial solution, the resulting characteristic equation can be solved numerically for the critical flow velocity, U. Extensive computer experimentation for obtaining the non-trivial solution of Eq.(19), for various pipe configurations, has demonstrated that the critical velocity can be multiple in some zones in the design space. This means that the eigenvalues cross each other, indicating multi-modal solutions (i.e. Bi- Tri- Quadri- modal solutions). Such a multiplicity introduces singularity of the eigenvalue derivatives with respect to the design variables, which does not allow the use of gradient methods. Therefore, it is necessary to formulate the optimization problem with respect to the critical velocity connected with two, three, or four simultaneous divergence modes. The present formulation employs multi-dimensional, non-gradient search techniques to find the required optimum solutions [2, 3]. This formulation requires only simple function evaluations without computing any derivatives for either the objective function or the design constraints. The additional constraints, which ought to be added to the optimization problem, are [18]:

$$U_1 \leq U_j \,, \quad j=2,3\ldots m. \tag{24}$$

where U_1 is the first eigenvalue representing the dimensionless critical flow velocity, U_j's are the subsequent higher eigenvalues and m is the assumed modality of the final optimum solution. All constraints are augmented with the objective function through penalty multiplier terms, and the number of active constraints at the optimum design point can automatically detect the actual modality of the problem. In the case of single mode optimization, none of the constraints become active at the optimal solution. It is noted that the total mass and length equality constraints can be used to eliminate some of the design variables, which help reducing the dimensionality of the optimization problem. The *MATLAB* optimization toolbox is a powerful tool that includes many routines for different types of optimization encompassing both unconstrained and constrained minimization algorithms [3]. One of its useful routines is named *"fmincon"* which finds the constrained minimum of an objective function of several variables. Figure 16 depicts the functional behavior of the dimensionless critical flow velocity, $U_{cr,1}$ augmented with the equality mass constraint, $M_s=1$. It is seen that the function is well behaved and continuous everywhere in the design space $(V_f\text{ -}L)_1$, except in the empty region located at the upper right of the whole domain, where the mass equality constraint is violated. The feasible domain is seen to be split by the baseline contours ($U_{cr}=\pi$) into two distinct zones. The one to the right encompasses the constrained global maxima, which is calculated to be $U_{cr}=3.2235$ at the optimal design point $(V_f\,,L)_{k=1,2}$ =(0.550, 0.80), (0.30, 0.20). Actually, each design point inside the feasible domain corresponds to different material properties as well as different stiffness and mass distributions, while maintaining the total structural mass constant. Figure 17 shows the developed isodiverts (lines of constant divergence velocity, $U_{cr,1}$) in the (V_{f1}-V_{f2}) design space. The equality mass constraint is violated in the first and third quadrants and the cross lines $V_{f1}=50\%$ and $V_{f2}=50\%$ represent the isodiverts of the baseline value π. For the case of a clamped-hinged (C/H), two-module pipe, the global maxima was calculated to be $(V_f\,,L)_{k=1,2}$ =(0.525, 0.875), (0.325, 0.125) at which $U_{cr,1}$=4.5645. Table 6 summarizes the attained optimal solutions for the different types of boundary conditions. Cases of combined material and thickness grading are also included, showing a truly and significant optimization gain for the different pipe configurations. More results indicated that for the case of *H/H* pipelines, good patterns must be symmetrical about the mid-span point. Therefore, it can be easier to cope with symmetrical configurations, which reduce computational efforts significantly, and the total number of variables to half. In this case, the boundary conditions become $w(0)=w''(0)=0$ and $w'(1/2)=w'''(1/2)=0$. For three-module H/H pipeline, the attained maximum value of the critical velocity was found to be 3.7955, occurring at the design point $(V_f, h, L)_k= (0.625,0.5,0.15625), (0.7,1.1375, 0.6875), (0.625, 0.5, 0.15625)$. This represents about 20.81% optimization gain above the baseline value π.

Support	$(V_f, h, L)_{k=1,2}$	$U_{cr,max}$
	Material grading only	
H/H	(0.550, 1.0, 0.800), (0.300, 1.0, 0.200)	3.2235
C/H	(0.525, 1.0, 0.875), (0.325, 1.0, 0.125)	4.5645
C/C	(0.675, 1.0, 0.125), (0.475, 1.0, 0.875)	6.3325
	Combined material & thickness grading	
H/H	(0.70, 1.0, 0.75), (0.65, 0.75, 0.25)	3.6235
C/H	(0.70, 0.95, 0.9), (0.50, 0.85, 0.10)	5.1355
C/C	(0.70, 1.0, 0.60), (0.65, 0.85, 0.40)	7.0965

Table 6. Optimal solutions for two-module pipelines

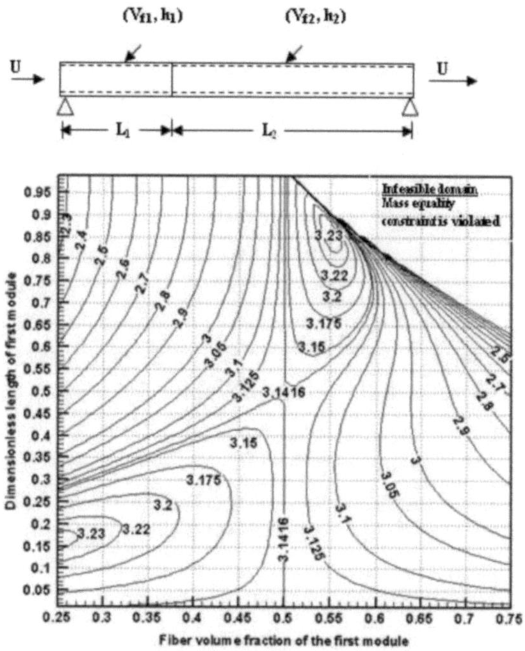

Figure 16. Effect of material grading on the critical flow velocity for a two-module, H/H pipe with constant total mass.

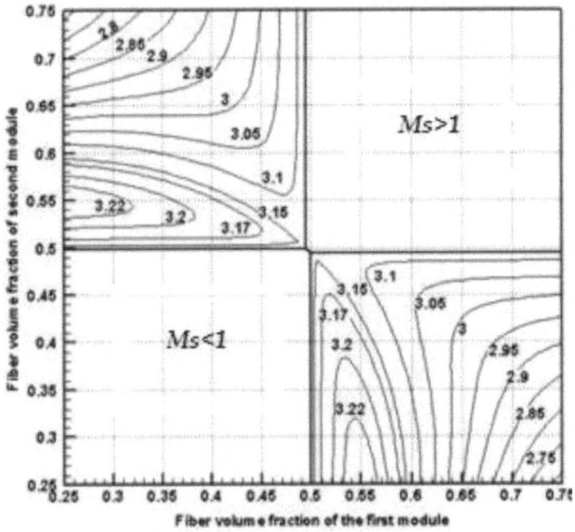

Figure 17. Isodiverts in the (V_{f1}-V_{f2}) design space for a two-module, *H*/*H* pipe.

7. Conclusion

As a major concern in producing efficient structures with enhanced properties and tailored response, this chapter presented appropriate design optimization models for improving performance and operational efficiency of different types of composite structural members. The concept of material grading was successfully applied by incorporating the distribution of the volume fractions of the composite material constituents in the mathematical formulation. Different optimization strategies have been addressed, including maximization of buckling stability of columns and cylindrical shells, natural frequencies of vibrating bars and critical flight speed of subsonic wings. Other stability problems concerning fluid-structure interaction has also been addressed. The general set of design variables encompasses volume fraction distribution, geometry and cross-sectional parameters. It has been shown that normalization of all quantities results in a naturally scaled objective functions, constraints and design variables, which is recommended when applying different optimization techniques. Piecewise models including multi-segment and multi-layered composite structures are implemented, where the optimized designs can be fabricated economically from any arbitrary number of uniform segments with material grading in a predetermined direction. Several design charts that are useful for direct determination of the optimal values of the design variables are given. It has been confirmed that the segment length is most significant design variable in the whole optimization process. Some investigators who apply finite elements have not recognized that the length of each element can be taken as a main design variable in the whole set of optimization variables. The results from the present approach reveal that piecewise grading of the material can be promising producing truly efficient designs with enhanced stability, dynamic and aeroelastic performance. Actually, the most economic structural design that will perform its intended function with adequate safety and durability requires much more than the procedures that have been described in this chapter. It is the author's wish that the results presented in this chapter will be compared and validated through other optimization techniques such as genetic algorithms or any appropriate global optimization algorithm. Further optimization studies must depend on a more accurate analysis of constructional cost. This combined with probability studies of load applications and materials variations, should contribute to further efficiency achievement. Much improved and economical designs for the main structural components may be obtained by considering multi-disciplinary design optimization, which allows designers to incorporate all relevant design objectives simultaneously. Finally, it is important to mention that, while FGM may serve as an excellent optimization and material tailoring tool, the ability to incorporate optimization techniques and solutions in practical design depend on the capacity to manufacture these materials to required specifications. Conventional techniques are often incapable of adequately addressing this issue. In conclusion, FGMs represent a rapidly developing area of science and engineering with numerous practical applications. The research needs in this area are uniquely numerous and diverse, but FGMs promise significant potential benefits that fully justify the necessary effort.

Author details

Karam Maalawi
National Research Centre, Mechanical Engineering Department, Cairo, Egypt

8. References

[1] Maalawi K., Badr M. (2009) Design Optimization of Mechanical Elements and Structures: A Review with Application. Journal of Applied Sciences Research 5(2): 221–231.

[2] Rao S. (2009) Engineering Optimization: Theory and Practice, 4th edition, John Wiley & Sons, ISBN: 978-0470183526, New York.

[3] Venkataraman P. (2009) Applied Optimization with MATLAB Programming, 2nd edition, John Wiley & Sons, ISBN: 978-0470084885, New York.

[4] Daniel I., Ishai O. (2006) Engineering Mechanics of Composite Materials, 2nd ed., Oxford Univ. Press, New York.

[5] Birman V., Byrd W. (2007) Modeling and Analysis of Functionally Graded Materials and Structures. Applied Mechanics Reviews, ASME 60: 195-216.

[6] Elishakoff I., Guede Z. (2004) Analytical Polynomial Solutions for Vibrating Axially Graded Beams. Journal of Mechanics and Advanced Materials and Structures, 11: 517-533.

[7] Elishakoff I., Endres J. (2005) Extension of Euler's Problem to Axially Graded Columns: Two Hundred and Sixty Years Later. Journal of Intelligent Material systems and Structures, 16(1): 77-83.

[8] Shi-Rong Li, Batra R. (2006) Buckling of Axially Compressed Thin Cylindrical Shells with Functionally Graded Middle Layer. Journal of Thin-Walled Structures, 44: 1039-1047.

[9] Qian L., Batra R. (2005) Design of Bidirectional Functionally Graded Plate for Optimal Natural Frequencies. Journal of Sound and Vibration, 280: 415-424.

[10] Goupee A., Vel S. (2006) Optimization of Natural Frequencies of Bidirectional Functionally Graded Beams. Journal of Structural and Multidisciplinary Optimization, 32(6): 473-484.

[11] Maalawi K. (2002) Buckling Optimization of Flexible Columns. International Journal of Solids and Structures, 39: 5865-5876.

[12] Maalawi K. (2009) Optimization of Elastic Columns using Axial Grading Concept. Engineering Structures, 31(12): 2922-2929.

[13] Maalawi K. (2011) Use of Material Grading for Enhanced Buckling Design of Thin-Walled Composite Rings/Long Cylinders under External Pressure. Composite Structures, 93(2): 351-359.

[14] Maalawi K. (2011) Functionally Graded Bars with Enhanced Dynamic Performance. Journal of Mechanics of Materials and Structures, 6(1-4): 377-393.

[15] Librescu L., Maalawi K. (2007) Material Grading for Improved Aeroelastic Stability in Composite Wings. Journal of Mechanics of Materials and Structures, 2(7): 1381-1394.

[16] Sheng G., Wang, X. (2010) Dynamic Characteristics of Fluid-Conveying Functionally Graded Cylindrical Shells under Mechanical and Thermal Loads. Composite Structures, 93(1): 162-170.

[17] Maalawi, K. and Ziada, M. (2002). On the Static Instability of Flexible Pipes Conveying Fluid. Journal of Fluids and Structures, 16(5): 685-690.

[18] Maalawi K., EL-Sayed H. (2011) Stability Optimization of Functionally Graded Pipes Conveying Fluid. Proceedings of the International Conference of Mechanical Engineering, World Academy of Science, Engineering and Technology, Paris, France, July 27-29, 2011, 178-183.

Chapter 3

Vibration Method in Stability Analysis of Planar Constrained Elastica

Jen-San Chen and Wei-Chia Ro

Additional information is available at the end of the chapter

http://dx.doi.org/10.5772/45806

1. Introduction

The primary goal of the research in constrained elastica is to understand the behavior of a thin elastic strip under end thrust when it is subject to lateral constraints. It finds applications in a variety of practical problems, such as in compliant foil journal bearing, corrugated fiberboard, deep drilling, structural core sandwich panel, sheet forming, non-woven fabrics manufacturing, and stent deployment procedure. By assuming small deformation, Feodosyev (1977) included the problem of a buckled beam constrained by a pair of parallel walls as an exercise for a university strength and material course. Vaillette and Adams (1983) derived a critical axial compressive force an infinitely long constrained elastica can support. Adams and Benson (1986) studied the post-buckling behavior of an elastic plate in a rigid channel. Chateau and Nguyen (1991) considered the effect of dry friction on the buckling of a constrained elastica. Adan et al. (1994) showed that when a column with initial imperfection positioned at a distance from a plane wall is subject to compression, contact zones may develop leading to buckling mode transition. Domokos et al. (1997), Holms et al. (1999), and Chai (1998, 2002) investigated the planar buckling patterns of an elastica constrained inside a pair of parallel plane walls. It was observed that both point contact and line contact with the constraint walls are possible. Kuru et al. (2000) studied the buckling behavior of drilling pipes in directional wells. Roman and Pocheau (1999, 2002) used an elastica model to investigate the post-buckling response of bilaterally constrained thin plates subject to a prescribed height reduction. Chen and Li (2007) and Lu and Chen (2008) studied the deformation of a planar elastica inside a circular channel with clearance. Denoel and Detournay (2011) proposed an Eulerian formulation of the constrained elastica problem.

The emphasis of these studies was placed on the static deformations of the constrained elastica. Very often, multiple equilibria under a specified set of loading condition are possible. Since only stable equilibrium configurations can exist in practice, there is a need to determine the stability of each of these equilibria in order to predict the behavior of the constrained elastica as the external load varies. For an unconstrained elastica, vibration method is commonly used to determine its stability; see Perkins (1990), Patricio et al. (1998), Santillan, et al. (2006), and Chen and Lin (2008). This conventional method, however, becomes useless in the case of constrained elastica.

The difficulty of the conventional vibration method arises from the existence of unilateral constraints. A unilateral constraint is capable of exerting compressive force onto the structure, but not tension. Mathematically, this type of constraints can be represented by a set of inequality equations. This poses challenges in determining the critical states of the loaded structure. In order to overcome this difficulty, the conventional stability analysis needs modification. In this chapter, we introduce a vibration method which is capable of determining the stability of a constrained elastica once the equilibrium configuration is known. The key of solving the vibration problem in constrained elastica is to take into account the sliding between the elastica and the space-fixed unilateral constraint during vibration.

In this chapter, we consider the vibration of an elastica constrained by a space-fixed point constraint. This particular constrained elastica problem is used to demonstrate the vibration method which is suitable to analyze the stability of a structure under unilateral constraint. In Section 2, we describe the studied problem in detail. In Section 3, we describe the static load-deflection relation. In Section 4, we introduce the theoretical formulation of the vibration method. In Section 5 an imperfect system when the point constraint is not at the mid-span is analyzed. In Section 6, several conclusions are summarized.

2. Problem description

Figure 1 shows an inextensible elastic strip with the right end fully clamped at a point B. On the left hand side there is a straight channel with an opening at point A. The distance between points A and B is L. Part of the strip is allowed to slide without friction and clearance inside the channel. A longitudinal pushing force F_A is applied at the left end of the strip inside the channel causing it to buckle in the domain of interest between points A and B. An xy-coordinate system is fixed at point A. A point H fixed at position $x = L/2$ and $y = h$ prevents the elastica from deforming freely after the elastica contacts point H.

The elastic strip is assumed to be straight and stress-free when F_A=0. The effect of gravity is ignored. The strip is uniform in all mechanical properties along its length. The length and the shape of the elastica in the domain of interest vary as the pushing force F_A increases. The boundary condition at point A may be called "partially clamped," by which we mean that the strip is allowed to slide freely through the opening A, while the lateral displacement and slope at A are fixed. The dashed and solid curves in Figure 1 represent two typical stages of the elastica deformation when F_A increases beyond the buckling load. The dashed curve is a symmetric deformation pattern before the elastica contacts the point constraint.

The solid curve represents an asymmetric deformation after the elastica contacts the point H. Other deformation patterns may also exist, which will be discussed later.

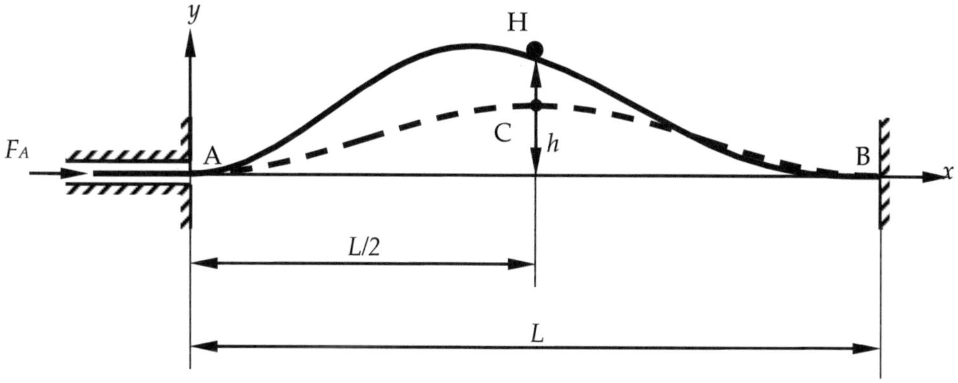

Figure 1. An elastica constrained by a space-fixed point at H. The dashed and solid curves represent typical symmetric and asymmetric deformations, respectively.

3. Load-deflection relation

The equilibrium equation at any point (x,y) of the buckled strip between points A and B, as shown in Figure 1, can be written as

$$EI\frac{d\theta}{ds} = -Q_A x - F_A y + M_A \tag{1}$$

Q_A and M_A are the internal shear force and bending moment, respectively, provided by the partial clamp at A. θ (positive when counter-clockwise) is the rotation angle of the strip at point (x,y). EI is the flexural rigidity of the elastic strip. s is the length of the strip measured from point A. For convenience we introduce the following dimensionless parameters (with asterisks):

$$(s^*, x^*, y^*) = \frac{(s,x,y)}{L},\ (Q_A^*, F_A^*) = \frac{L^2}{4\pi^2 EI}(Q_A, F_A),\ M_A^* = \frac{L}{4\pi^2 EI} M_A$$

$$t^* = \frac{1}{L^2}\sqrt{\frac{EI}{\mu}}\, t,\ \omega^* = L^2\sqrt{\frac{\mu}{EI}}\,\omega$$

μ is the mass per unit length of the elastica. t is time and ω is a circular natural frequency, which will be discussed in the dynamic analysis later. After substituting the above relations into Equation (1), and dropping all the superposed asterisks thereafter for simplicity, we obtain the dimensionless equilibrium equation

$$\frac{d\theta}{ds} = 4\pi^2\left(M_A - Q_A x - F_A y\right) \tag{2}$$

The method of static analysis can be found in Chen and Ro (2010). In this section we introduce several deformation patterns of the constrained elastica. All the physical quantities described henceforth are dimensionless.

The length of the elastica being pushed in through the opening is $\Delta l = l - 1$, where l is the dimensionless length of the elastica between points A to B. Figure 2 shows the relation between the edge thrust F_A and the length increment Δl. The height of the point constraint h is 0.03. The dashed and solid curves in this load-deflection diagram represent unstable and stable configurations, respectively. The method used in determining the stability of the static deformation will be described in detail in Section 4.

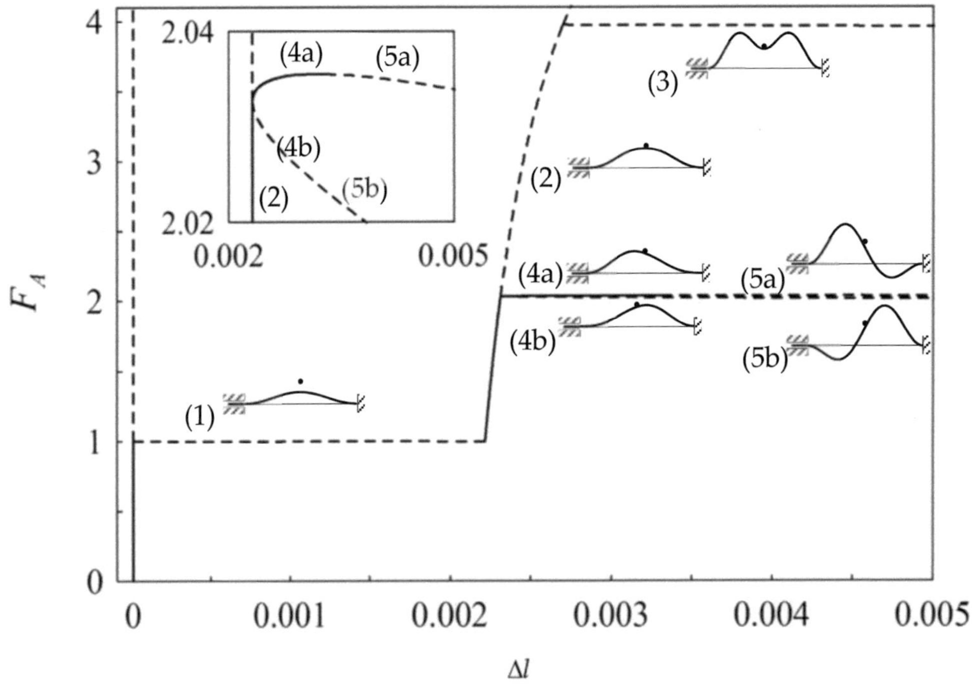

Figure 2. Load-deflection curves for h=0.03. The solid and dashed curves represent stable and unstable deformations, respectively.

The symmetric deformation before contact occurs is called deformation (1), whose locus starts at (F_A, Δl) = (1,0) and ends at (0.99668, 0.0022188). The slope of this load-deflection curve is slightly negative. After the middle point C of the elastica touches the point constraint H, the deformation pattern initially remains symmetric, called deformation (2). The load-deflection curve of deformation (2) starts at (F_A, Δl)=(0.99668,0.0022188) and ends at (3.97314, 0.0026985). It is noted that the lower part of this load-deflection curve up to

(2.03268, 0.0023129) is solid and the upper part is dashed. At the point separating the solid and dashed parts, a symmetry-breaking bifurcation occurs and the elastica evolves to a pair of asymmetric deformations 4(a) and 4(b). As the pushing force continues to increase, it is natural to envision that the elastica may evolve to an "M" shape, i.e., there exist two inflection points in each half of the span, called deformation (3). The load-deflection curve corresponding to deformation (3) starts at $(F_A, \Delta l)$=(3.97314, 0.0026985) and continues beyond the range of Figure 2. The slope of this curve is slightly negative.

At point $(F_A, \Delta l)$=(2.03268, 0.0023129), the symmetric deformation (2) described previously may bifurcate to a pair of asymmetric deformations 4(a) and 4(b). Both of (4a) and (4b) have one inflection point on each of the half span separated by the point constraint. Both (4a) and (4b) start at $(F_A, \Delta l)$=(2.03268, 0.0023129), while (4a) ends at (2.03545, 0.0028996) and (4b) ends at (2.02600 0.0028996). The slopes of curves (4a) and (4b) are positive and negative throughout, respectively. Stability analysis later indicates that deformation (4b) is unstable while (4a) is stable. As Δl increases further, the pair of deformations (4a) and (4b) evolve to a pair of asymmetric deformations (5a) and (5b). Deformation (5a) has two inflection points on the left span and one inflection point on the right span. On the other hand, deformation (5b) has two inflection points on the right span and one inflection point on the left span. The slope of load-deflection curve of (5b) is negative throughout. On the other hand, the curve of (5a) is of convex shape with the top being at point $(F_A, \Delta l)$ =(2.03554, 0.0031711), which is very close to the end of curve of deformation (4a). Deformation (5a) is stable first until the curve reaches the top at $(F_A, \Delta l)$ =(2.03554, 0.0031711). At this critical point, the elastica will jump to another configuration. The load-deflection curves near the symmetry-breaking point are magnified and shown in the inset of Figure 2.

The theoretical load-deflection curves shown in Figure 2 give us a mental picture how the elastica evolves as the pushing force F_A increases. First of all, the elastica remains still when F_A is smaller than 1, the Euler buckling load. As soon as F_A reaches 1, the elastica jumps to symmetric deformation (2) in contact with the point constraint. As F_A increases, a symmetry-breaking bifurcation occurs and the elastica evolves to asymmetric deformation (4a) first and smoothly to (5a). As F_A continues to increase up to a certain value, a second jump occurs. Following this jump the elastica will eventually settle to a self-contact configuration. This self-contact configuration requires a length increment Δl over 8, which is well beyond the range of Figure 2. The above scenario has been verified experimentally in Chen and Ro (2010). In the next section we describe the vibration method used to determine the stability of the static deformations.

4. Vibration and stability analyses

4.1. Lagrangian and Eulerian descriptions

As mentioned above, the deformation patterns discussed in Section 3 may not necessarily be stable. If the deformation is unstable, then it can not be realized in practice. In order to study the vibration and stability properties of the elastica, we first derive the equations of motion of a small element *ds* supported by the point constraint, as shown in Figure 3.

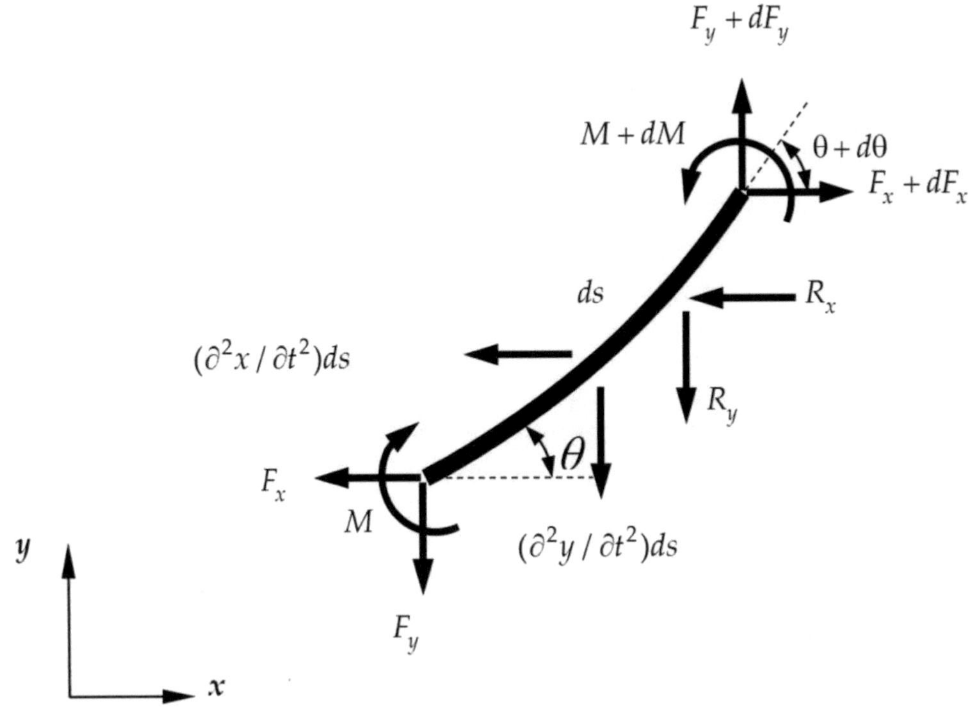

Figure 3. The free body diagram of a small element *ds* constrained by the space-fixed point.

The geometrical relations between x, y, and θ are

$$\frac{\partial x(s,t)}{\partial s} = \cos\theta(s,t) \tag{3}$$

$$\frac{\partial y(s,t)}{\partial s} = \sin\theta(s,t) \tag{4}$$

The balance of moment and forces in the x- and y-directions results in

$$\frac{\partial M(s,t)}{\partial s} = F_x(s,t)\sin\theta(s,t) - F_y(s,t)\cos\theta(s,t) \tag{5}$$

$$\frac{\partial F_x(s,t)}{\partial s} - R_x(t)\delta\left(s - l_1 - \eta_1(t)\right) = \frac{1}{4\pi^2}\frac{\partial^2 x(s,t)}{\partial t^2} \tag{6}$$

$$\frac{\partial F_y(s,t)}{\partial s} - R_y(t)\delta\left(s - l_1 - \eta_1(t)\right) = \frac{1}{4\pi^2}\frac{\partial^2 y(s,t)}{\partial t^2} \tag{7}$$

$F_x(s,t)$ and $F_y(s,t)$ are the internal forces in the x- and y-directions. The moment-curvature relation of the Euler-Bernoulli beam model is

$$\frac{\partial\theta(s,t)}{\partial s} = 4\pi^2 M(s,t) \tag{8}$$

The readers are reminded that the functions x, y, θ, M, F_x, and F_y in Equations (3)-(8) are dimensionless and are all written explicitly in terms of s and t for clarity. These six equations can be called the Lagrangian version of the governing equations because a material element ds at location s is isolated as the free body. s may be called the Lagrangian coordinate of a point on the elastica. It is noted that s=0, l_1, and l represent the material points at the left end, the contact point, and the right end, respectively, when the elastica is in equilibrium. During vibration, the elastica may "slide" on the point constraint. As a consequence, the contact point on the elastica may change from s=l_1 to $s = l_1 + \eta_1$, where $\eta_1(t)$ is a small number. This change of contact point is reflected in Equations (6) and (7). $R_x(t)$ and $R_y(t)$ are the x- and y-component forces exerted by the point constraint on the elastica during vibration. $\delta(\cdot)$ is the dimensionless Dirac delta function.

We denote the static solutions of Equations (3)-(8) as $x_e(s)$, $y_e(s)$, $\theta_e(s)$, $M_e(s)$, $F_{xe}(s)$, and $F_{ye}(s)$. It is assumed that these static solutions are known. In the case when contact occurs, the relations between F_{xe}, F_{ye} and F_A, Q_A are

$$F_{xe}(s) = -F_A + R_{xe}H(s - l_1) \tag{9}$$

$$F_{ye}(s) = Q_A + R_{ye}H(s - l_1) \tag{10}$$

H is the Heaviside step function. During vibration, the function $F_y(s,t)$ may be regarded as the superposition of $F_{ye}(s)$ in Equation (10) and a small harmonic perturbation, expressed mathematically as

$$F_y(s,t) = F_{ye}(s) + \left[R_{ye}(H(s - l_1 - \eta_1) - H(s - l_1))\right] +$$

$$\left[F_{yd}(s - \eta_1) + R_{yd}H(s - l_1 - \eta_1)\right]\sin\omega t \tag{11}$$

A variable with subscript "d" represents a small perturbation of its static counterpart with subscript "e."

Figure 4 is a graphical interpretation of Equation (11). The solid step lines in Figure 4(a) represent $F_{ye}(s)$. After sliding occurs the cross-hatched area disappears and the contact point moves from $s = l_1$ to $s = l_1 + \eta_1$. This shift of the contact point from s=l_1 to $s = l_1 + \eta_1$ is represented by the first bracket on the right hand side of Equation (11). Figure 4(b) shows the superposition of reactive point force R_{yd} to the right of the new contact point. This action is represented by the second term in the second bracket on the right hand side of Equation (11). Finally, Figure 4(c) shows the superposition of $\hat{F}_{yd}$. The step dashed curve in Figure 4(c) represents the final $F_y(s,t)$ in Equation (11).

After defining a new variable ε as

$$\varepsilon = s - \eta_1, \tag{12}$$

Equation (11) can be rewritten as

$$\hat{F}_y(\varepsilon, t) = F_{ye}(\varepsilon) + \left[F_{yd}(\varepsilon) + R_{yd} H(\varepsilon - l_1)\right]\sin\omega t \tag{13}$$

where $\hat{F}_y(\varepsilon,t) = F_y(\varepsilon + \eta_1, t)$. Apparently, $\hat{F}_y$ and F_y are two different functions. It is noted that $F_{ye}(\varepsilon)$ is the static solution as obtained from the static analysis, except that the independent variable s is replaced by ε. Similarly, the other perturbed functions may be written as

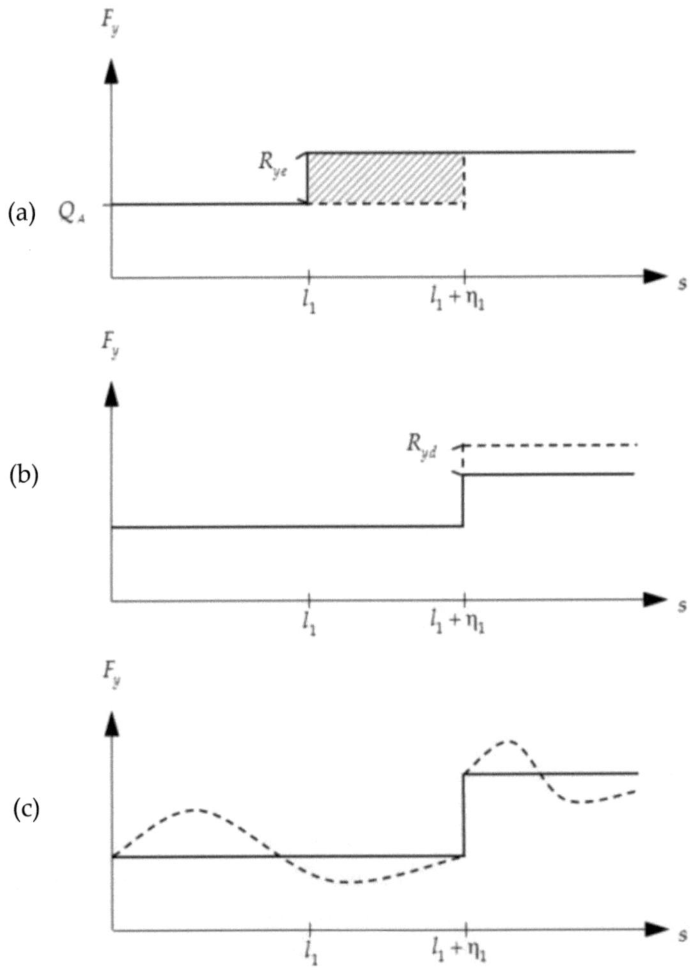

Figure 4. Schematic diagram to demonstrate the perturbation of $F_y(s,t)$, refer to Equation (11).

$$\hat{F}_x(\varepsilon,t) = F_{xe}(\varepsilon) + \left[F_{xd}(\varepsilon) + R_{xd}H(\varepsilon - l_1)\right]\sin\omega t \tag{14}$$

$$\hat{x}(\varepsilon,t) = x_e(\varepsilon) + x_d(\varepsilon)\sin\omega t \tag{15}$$

$$\hat{y}(\varepsilon,t) = y_e(\varepsilon) + y_d(\varepsilon)\sin\omega t \tag{16}$$

$$\hat{\theta}(\varepsilon,t) = \theta_e(\varepsilon) + \theta_d(\varepsilon)\sin\omega t \tag{17}$$

$$\hat{M}(\varepsilon,t) = M_e(\varepsilon) + M_d(\varepsilon)\sin\omega t \tag{18}$$

ε defined in Equation (12) may be called the Eulerian coordinate of a point on the elastica. $\varepsilon = l_1$ represents the point of the elastica passing through the point constraint at any instance during vibration. It can be a different material point at a different instant. The physical meaning of $\varepsilon = l_1$ is like fixing a control window at the point constraint. Therefore, we call this type of description an Eulerian one.

By noting that $\frac{\partial \varepsilon}{\partial s} = 1$, the Lagrangian version of the governing equations (3)-(8) can now be transformed into the Eulerian version as

$$\frac{\partial \hat{x}(\varepsilon,t)}{\partial \varepsilon} = \cos\hat{\theta}(\varepsilon,t) \tag{19}$$

$$\frac{\partial \hat{y}(\varepsilon,t)}{\partial \varepsilon} = \sin\hat{\theta}(\varepsilon,t) \tag{20}$$

$$\frac{\partial \hat{M}(\varepsilon,t)}{\partial \varepsilon} = \hat{F}_x(\varepsilon,t)\sin\hat{\theta}(\varepsilon,t) - \hat{F}_y(\varepsilon,t)\cos\hat{\theta}(\varepsilon,t) \tag{21}$$

$$\frac{\partial \hat{F}_x(\varepsilon,t)}{\partial \varepsilon} - R_x(t)\delta(\varepsilon - l_1) = \frac{1}{4\pi^2}\frac{\partial^2 \hat{x}(\varepsilon,t)}{\partial t^2} \tag{22}$$

$$\frac{\partial \hat{F}_y(\varepsilon,t)}{\partial \varepsilon} - R_y(t)\delta(\varepsilon - l_1) = \frac{1}{4\pi^2}\frac{\partial^2 \hat{y}(\varepsilon,t)}{\partial t^2} \tag{23}$$

$$\frac{\partial \hat{\theta}(\varepsilon,t)}{\partial \varepsilon} = 4\pi^2 \hat{M}(\varepsilon,t) \tag{24}$$

By substituting Equations (13)-(18), together with the relations

$$R_x(t) = R_{xe} + R_{xd}\sin\omega t \tag{25}$$

$$R_y(t) = R_{ye} + R_{yd}\sin\omega t \tag{26}$$

$$\eta_1(t) = \eta_{1d} \sin \omega t \tag{27}$$

into Equations (19)-(24) and ignoring the higher-order terms, we arrive at the following linear equations for the six functions $x_d(\varepsilon)$, $y_d(\varepsilon)$, $\theta_d(\varepsilon)$, $M_d(\varepsilon)$, $F_{xd}(\varepsilon)$, and $F_{yd}(\varepsilon)$:

$$\frac{dx_d(\varepsilon)}{d\varepsilon} = -\theta_d(\varepsilon)\sin\theta_e(\varepsilon) \tag{28}$$

$$\frac{dy_d(\varepsilon)}{d\varepsilon} = \theta_d(\varepsilon)\cos\theta_e(\varepsilon) \tag{29}$$

$$\frac{d\theta_d(\varepsilon)}{d\varepsilon} = 4\pi^2 M_d(\varepsilon) \tag{30}$$

$$\frac{dM_d(\varepsilon)}{d\varepsilon} = \left[F_{xe}(\varepsilon)\theta_d(\varepsilon) - F_{yd}(\varepsilon) - R_{yd}H\left(\varepsilon - l_1\right)\right]\cos\theta_e(\varepsilon)$$

$$+\left[F_{ye}(\varepsilon)\theta_d(\varepsilon) + F_{xd}(\varepsilon) + R_{xd}H\left(\varepsilon - l_1\right)\right]\sin\theta_e(\varepsilon) \tag{31}$$

$$\frac{dF_{xd}(\varepsilon)}{d\varepsilon} = -\frac{1}{4\pi^2}\omega^2\left[x_d(\varepsilon) - \cos\theta_e(\varepsilon)\eta_{1d}\right] \tag{32}$$

$$\frac{dF_{yd}(\varepsilon)}{d\varepsilon} = -\frac{1}{4\pi^2}\omega^2\left[y_d(\varepsilon) - \sin\theta_e(\varepsilon)\eta_{1d}\right] \tag{33}$$

4.2. Boundary conditions

The exact boundary conditions at the fixed end B are

$$x(s,t)\big|_{s=l} = \hat{x}(\varepsilon,t)\big|_{\varepsilon=l-\eta_1} = 1 \tag{34}$$

$$y(s,t)\big|_{s=l} = \hat{y}(\varepsilon,t)\big|_{\varepsilon=l-\eta_1} = 0 \tag{35}$$

$$\theta(s,t)\big|_{s=l} = \hat{\theta}(\varepsilon,t)\big|_{\varepsilon=l-\eta_1} = 0 \tag{36}$$

These boundary conditions can be linearized as follows. Take Equation (34) as an example. By using Equation (15), we can rewrite (34) into

$$x_e(\varepsilon)\big|_{\varepsilon=l-\eta_1} + x_d(\varepsilon)\big|_{\varepsilon=l-\eta_1}\sin\omega t = 1 \tag{37}$$

Both $x_e(\varepsilon)\big|_{\varepsilon=l-\eta_1}$ and $x_d(\varepsilon)\big|_{\varepsilon=l-\eta_1}$ in Equation (37) can be expanded as a Taylor series with respect to $\varepsilon = l$. After ignoring the higher-order terms, Equation (37) can be linearized to

$$x_d(\varepsilon)\big|_{\varepsilon=l} = \eta_{1d} \tag{38}$$

Similarly, the boundary conditions (35)-(36) can be linearized to

$$y_d(\varepsilon)\big|_{\varepsilon=l} = 0 \tag{39}$$

$$\theta_d(\varepsilon)\big|_{\varepsilon=l} = 4\pi^2 M_e(\varepsilon)\big|_{\varepsilon=l}\,\eta_{1d} \tag{40}$$

The boundary condition at the left end A is more complicated. We denote the material point on the strip right at the opening A of the channel as point A′ when the elastica is in equilibrium, as shown in Figure 5(a). Since the strip is under a constant pushing force at the left end, A′ will retreat into and protrude out of the channel when the elastica vibrates, as shown in Figures 5(b) and 5(c). We denote this small length of movement as

$$\eta_0(t) = \eta_{0d}\sin\omega t \tag{41}$$

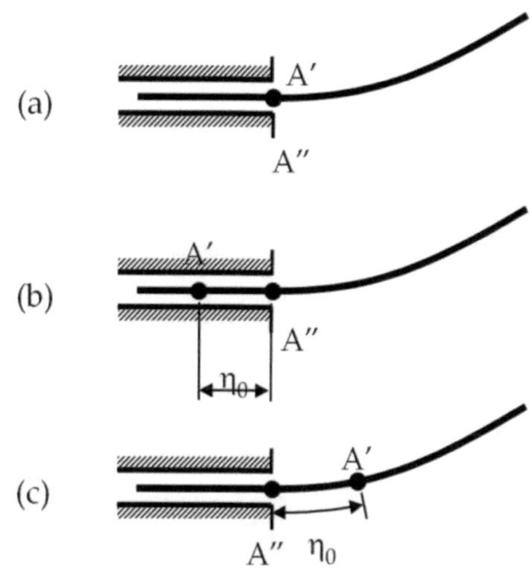

Figure 5. The boundary conditions at the opening A of the feeding channel. (a) In equilibrium position, the material point A′ coincides with point A. When the elastica vibrates, the material point A′ (b) retreats in and (c) protrudes out of the channel.

The condition of zero slope at opening A requires that

$$\theta(s,t)\big|_{s=\eta_0} = \hat{\theta}(\varepsilon,t)\big|_{\varepsilon=\eta_0-\eta_1} = 0 \tag{42}$$

Following the similar linearization procedure as at point B, we can linearize boundary condition (42) to the form

$$\theta_d(\varepsilon)\big|_{\varepsilon=0} = -4\pi^2 M_e(\varepsilon)\big|_{\varepsilon=0}(\eta_{0d} - \eta_{1d}) \tag{43}$$

Similarly, we can derive

$$x_d(\varepsilon)\big|_{\varepsilon=0} = -(\eta_{0d} - \eta_{1d}) \tag{44}$$

$$y_d(\varepsilon)\big|_{\varepsilon=0} = 0 \tag{45}$$

$$F_{xd}(\varepsilon)\big|_{\varepsilon=0} = 0 \tag{46}$$

Finally, Equations (43) and (44) may be combined as

$$\theta_d(\varepsilon)\big|_{\varepsilon=0} = 4\pi^2 M_e(\varepsilon)\big|_{\varepsilon=0}\, x_d(\varepsilon)\big|_{\varepsilon=0} \tag{47}$$

The three equations (45)-(47) are the linearized boundary conditions at point A.

4.3. Constraint equations

When contact occurs, it is required that the elastica always passes through the point constraint. Mathematically, this condition can be written as

$$x(s,t)\big|_{s=l_1+\eta_1} = \hat{x}(\varepsilon,t)\big|_{\varepsilon=l_1} = 0.5 \tag{48}$$

$$y(s,t)\big|_{s=l_1+\eta_1} = \hat{y}(\varepsilon,t)\big|_{\varepsilon=l_1} = h \tag{49}$$

After using Equations (15)-(16), Equations (48)-(49) can be rewritten as

$$x_d(\varepsilon)\big|_{\varepsilon=l_1} = 0 \tag{50}$$

$$y_d(\varepsilon)\big|_{\varepsilon=l_1} = 0 \tag{51}$$

We also require that the dynamic reactive force must be always normal to the elastica at the point constraint, or mathematically,

$$R_x \cos\hat{\theta}(\varepsilon,t) + R_y \sin\hat{\theta}(\varepsilon,t) = 0 \quad \text{at } \varepsilon = l_1 \tag{52}$$

After using Equations (17), (25)-(26) and neglecting higher-order terms, Equation (52) can be linearized to

$$\left[R_{ye}\theta_d(\varepsilon) + R_{xd}\right]\cos\theta_e(\varepsilon) + \left[-R_{xe}\theta_d(\varepsilon) + R_{yd}\right]\sin\theta_e(\varepsilon) = 0 \quad \text{at } \varepsilon = l_1 \tag{53}$$

Equations (50), (51), and (53) are the three constraint equations.

4.4. Solution method

In summary, the six linearized differential equations (28)-(33), six boundary conditions (38)-(40), (45)-(47), and three constraint equations (50)-(51) and (53) admit nontrivial solutions only when ω is equal to an eigenvalue of the system of equations. The unknowns to be found are the six functions $x_d(\varepsilon)$, $y_d(\varepsilon)$, $\theta_d(\varepsilon)$, $M_d(\varepsilon)$, $F_{xd}(\varepsilon)$, $F_{yd}(\varepsilon)$, the amplitude of sliding at the point constraint η_{1d}, and the two dynamic constraint reactions R_{xd} and R_{yd}. It is noted that ω in Equations (32)-(33) only appears in the form of ω^2. Therefore, if the characteristic value ω^2 is positive, the corresponding mode is stable with natural frequency ω. On the other hand, the equilibrium configuration is unstable if ω^2 is negative.

A shooting method is used to solve for the characteristic value ω^2. Since the linear vibration mode shape is independent of the amplitude, we can set $M_d(\varepsilon)\big|_{\varepsilon=0}=1$. After guessing six variables $x_d(\varepsilon)\big|_{\varepsilon=0}$, $F_{yd}(\varepsilon)\big|_{\varepsilon=0}$, R_{xd}, R_{yd}, η_{1d}, and ω^2, we can integrate the homogeneous equations (28)-(33) like an initial value problem all the way from $\varepsilon=0$ to $\varepsilon=l$. The three boundary conditions (38)-(40) at $\varepsilon=l$ and the three constraint equations (50)-(51) and (53) at $\varepsilon=l_1$ are used to check the accuracy of the guesses. If the guesses are not satisfactory, a new set of guesses is adopted. The stability of the deformations in Figure 2 is determined in this manner. It is noted that sometimes the term $M_d(\varepsilon)\big|_{\varepsilon=0}$ of a mode shape happens to be zero. In such a case, the assumption $M_d(\varepsilon)\big|_{\varepsilon=0}=1$ will yield no solution. When this situation occurs, a different variable is set as 1 in the shooting method; for instance, $x_d(\varepsilon)\big|_{\varepsilon=0}=1$.

Figure 6 shows the ω^2 of the first two modes as functions of the end force F_A for deformation (2). The ω^2 of the first mode becomes negative when F_A reaches 2.03268, at which the symmetry-breaking bifurcation occurs. It is noted that the ω^2 of the second mode

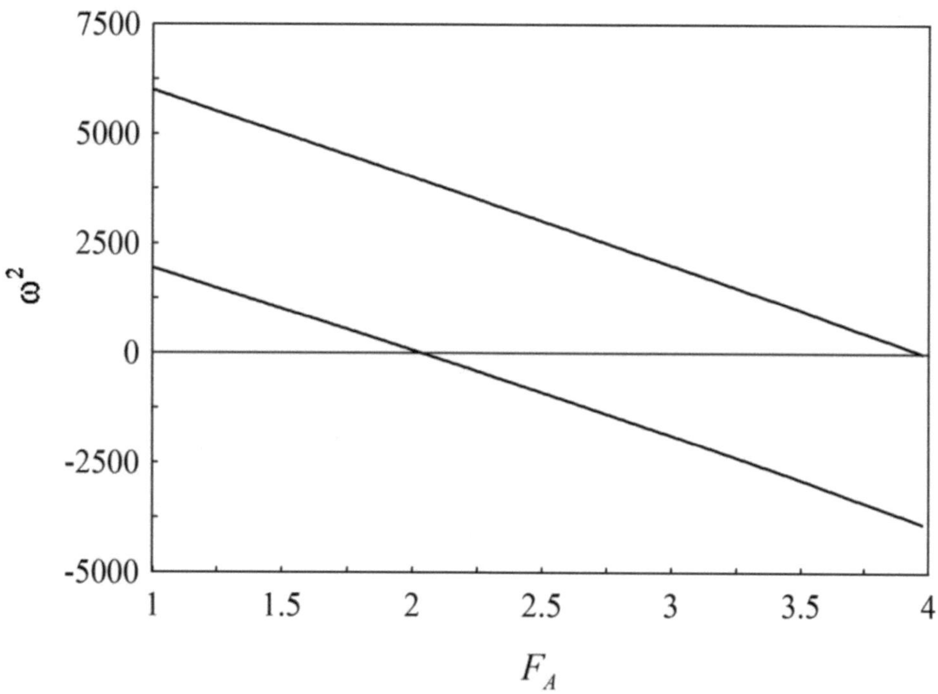

Figure 6. ω^2 of the first two modes as functions of the end force F_A for deformation (2).

becomes negative when F_A reaches 3.97314. This happens to be the point at which deformation (3) begins to appear.

The first two mode shapes when F_A=1.5 are depicted in Figure 7. The solid and dashed curves represent the static and the vibrating mode shapes of the constrained elastica, respectively. The first mode (a) is asymmetric and the second mode (b) is symmetric. To examine whether sliding at the contact point occurs we examine the values of η_{0d} and η_{1d} of each mode. For the asymmetric mode we found that the ratio η_{0d}/η_{1d} is approximately 1:400. This means that during vibration the protruding and retreating of the elastica at the channel opening A is negligible compared to the sliding at the contact point. In other words, the elastica length within the domain of interest is almost constant during vibration. This can also be observed from Figure 7(a). For the symmetric mode, we found that the ratio η_{0d}/η_{1d} is approximately 2:1. Therefore, sliding at the contact point still occurs. From Figure 7(b) we can observe that the lengths of the elastica on both sides of the point constraint increase (or decrease during the other half of the period) the same amount. Since the elastica is inextensible, the protruding η_{0d} at the channel opening has to be twice the amount of the sliding η_{1d} at the contact point. It is noted that a vibration analysis of a constrained structure will cause an erroneous result if sliding at the constraint is neglected.

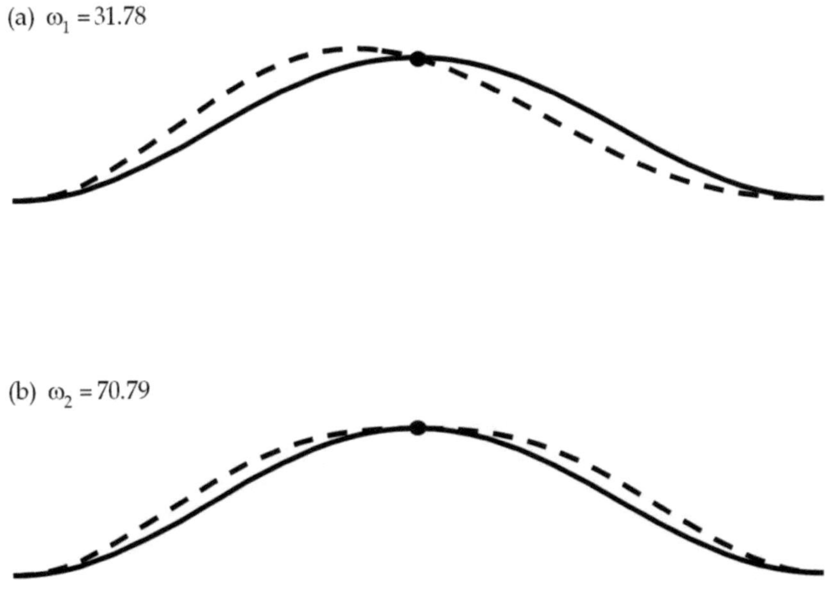

Figure 7. The first two mode shapes of the constrained elastica when F_A=1.5.

It is noted that the geometric conditions (50)-(51) at the point constraint are exact. Therefore, the vibrating elastica always passes through the point constraint no matter how large the vibration amplitude is. On the other hand, the boundary conditions at points A and B used in the calculation have been linearized from the exact boundary conditions. Therefore, the mode shapes do not necessarily satisfy the exact boundary conditions at points A and B. This may become obvious when the amplitude of vibration is increased dramatically.

Figures 8(a) and 8(b) show the lowest ω^2, i.e., ω_1^2, as a function of the length increment Δl for deformations (4a) and (5a), respectively. ω_1^2 of deformation (4a) is always positive. On the other hand, ω_1^2 of deformation (5a) becomes negative when Δl reaches 0.0031711. This is the point at which the load-deflection locus of deformation (5a) reaches its top in Figure 2. The mode shapes of deformations (4a) and (5a) are also depicted in the graphs. Similarly, the ω_1^2 of deformations (4b) and (5b) are plotted in Figures 9(a) and 9(b). Since ω_1^2 is always

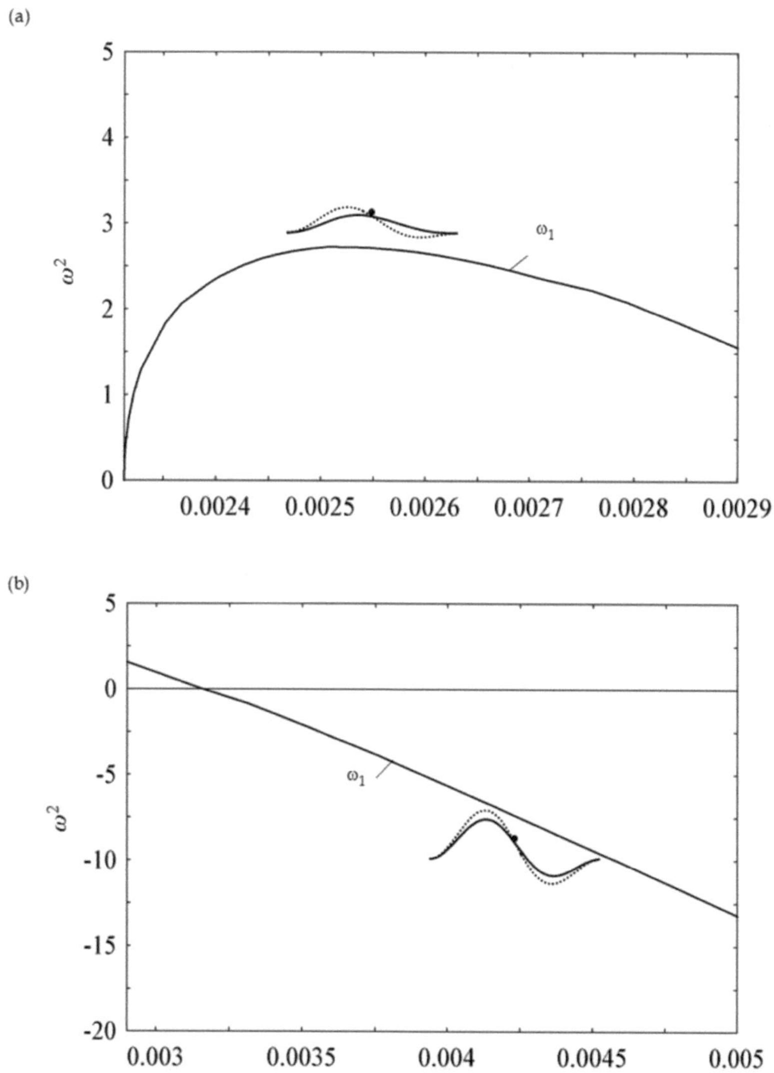

Figure 8. ω_1^2 as a function of the length increment Δl for deformations (a) (4a) and (b) (5a). The ω_1^2 of deformation (5a) becomes negative when Δl reaches 0.0031711.

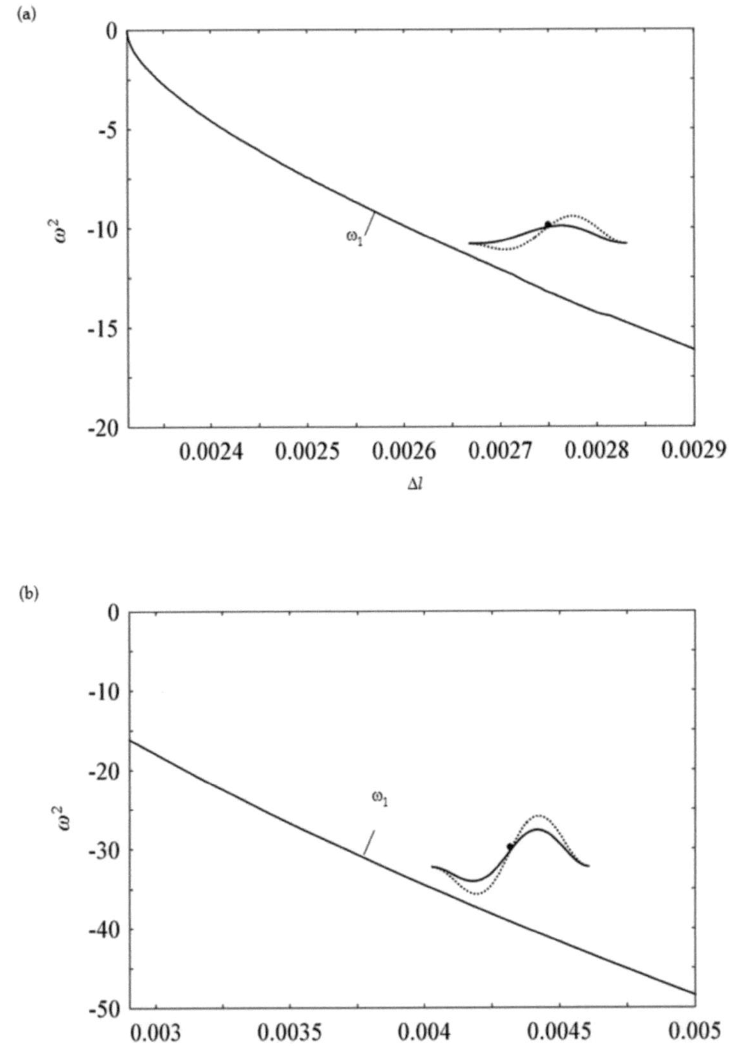

Figure 9. ω_1^2 as a function of the length increment Δl for deformations (a) (4b) and (b) (5b). The ω_1^2 of deformations (4b) and (5b) is always negative.

negative along the loci of deformations (4b) and (5b), we conclude that deformations (4b) and (5b) are unstable. From these stability analyses, it is concluded that after the symmetry-breaking bifurcation of deformation (2), the elastica branches to deformation 4(a) and continue to deform along locus (5a). After the Δl reaches 0.0031711, the elastica will jump to a remote self-contact configuration beyond the range of Figure 2.

5. Analysis of an imperfect system

The point constraint H in Figure 1 is at the middle between the two ends A and B. In practice, it is very difficult to place the point constraint accurately at the center. Instead, it is almost inevitable that the point constraint may be off the center somewhat. Figure 10 shows the configuration when the point constraint H (black dot) is at a distance δ_H to the left of the ideal center (open circle). If the point constraint is on the right, δ_H is considered to be negative.

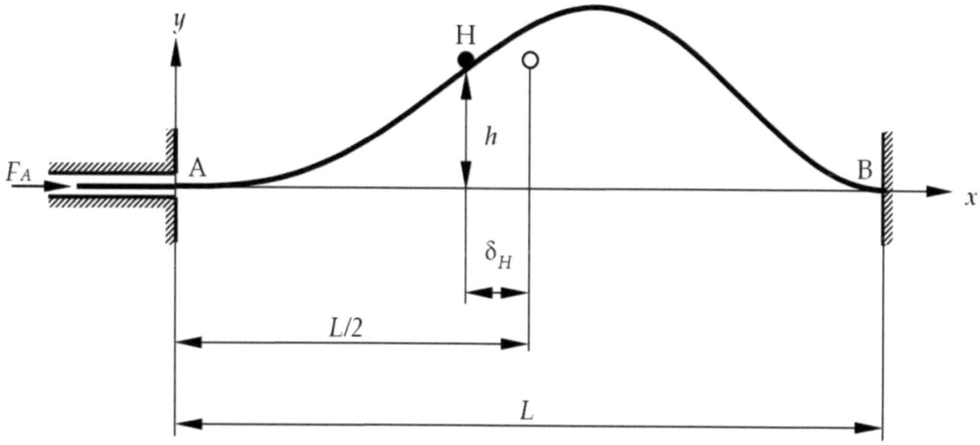

Figure 10. The point constraint H (black dot) is at a distance δ_H to the left of the ideal center (open circle). If H is on the right, δ_H is considered to be negative.

In Figure 11 we describe the change of the load-deflection relation when δ_H is increased from 0 (ideal case) to 0.01 and 0.05. The height h remains to be 0.03. Focus is placed on how the offset affects the symmetry-breaking bifurcation when deformation (2) branches into asymmetric deformations 4(a) and 4(b) in Figure 2. It is observed that the sharp corner at the bifurcation point degenerates into two smooth curves, called deformations 6(a) and 6(b) in Figure 11. Both deformations 6(a) and 6(b) are asymmetric. The tops of deformation 6(a) and 6(b) are to the left and right, respectively, of the point constraint H. Deformation 6(b) is always unstable. Deformation 6(a) for δ_H =0.01, on the other hand, is stable before point (F_A , Δl)=(1.974, 0.00552). Figure 12 shows the ω_1^2 along the locus of deformation 6(a) when δ_H =0.01. It is shown that ω^2 becomes negative when Δl =0.00552. For a larger δ_H =0.05, deformation 6(b) is stable throughout the range in Figure 11.

Inspecting Figure 11 reveals something unusual about the degeneration of the symmetry-breaking bifurcation due to the offset of the point constraint. For the ideal case with δ_H =0, part of upper branch (deformations 4(a) and (5a)) is stable until it reaches a maximum. On the other hand, the lower branch (deformations 4(b) and (5b)) is always unstable. When δ_H increases from 0 to 0.01, part of the lower branch is stable until it reaches a maximum at Δl =0.00552. On the other hand, the upper branch is always unstable. It is not clear how the load-deflection curves evolve as δ_H varies.

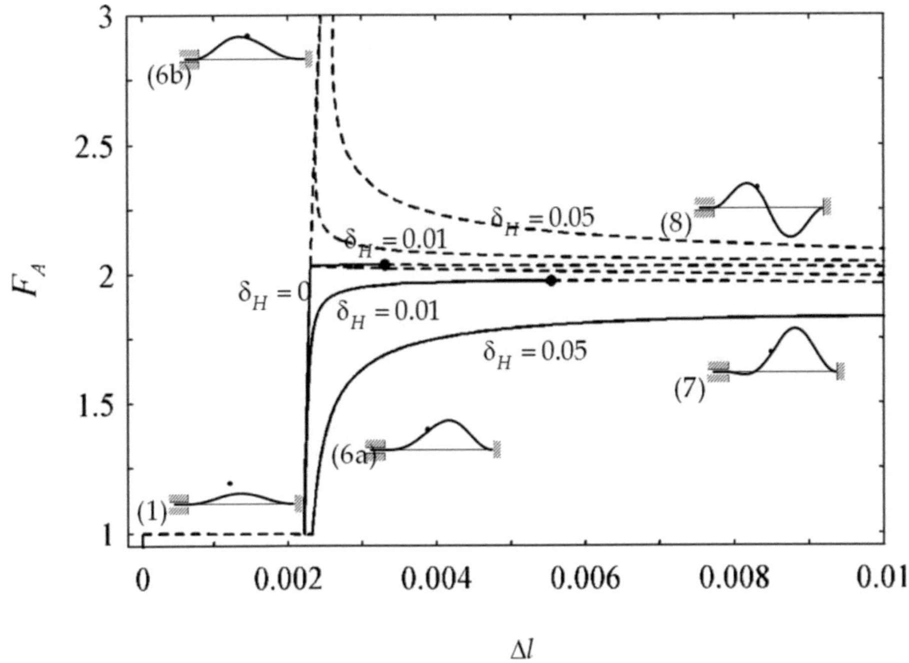

Figure 11. Load-deflection curves for h=0.03 and δ_H =0 (ideal case), 0.01, and 0.05, respectively. The solid and dashed curves represent stable and unstable deformations, respectively.

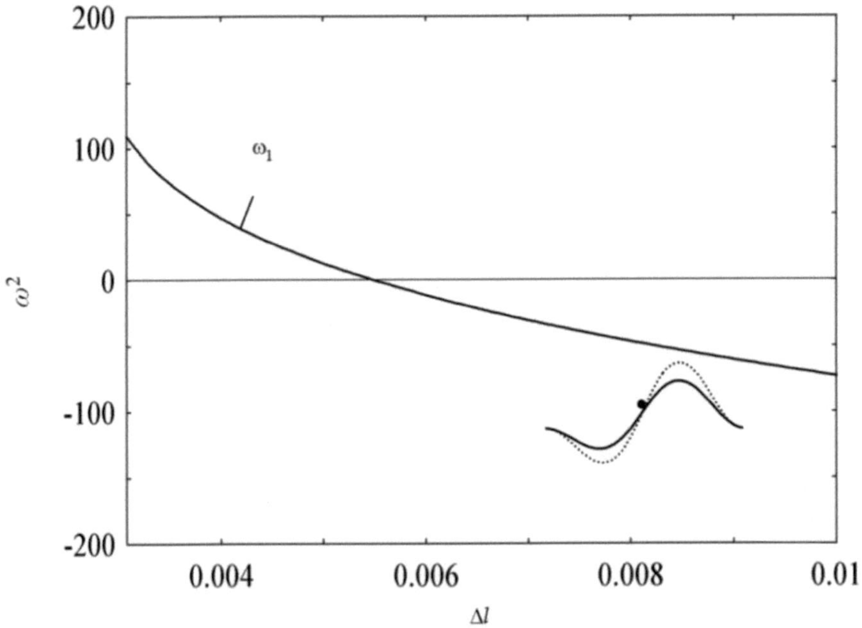

Figure 12. ω^2 of the first mode as functions of Δl for deformation (6) in Figure 11.

In order to answer this question, we plot the load-deflection curves when δ_H varies with smaller increment. Figure 13(a) shows the load-deflection curves when δ_H =0, 5×10^{-5}, and 2×10^{-4}, respectively. For the case when δ_H =0, part of the upper branch is stable, while the

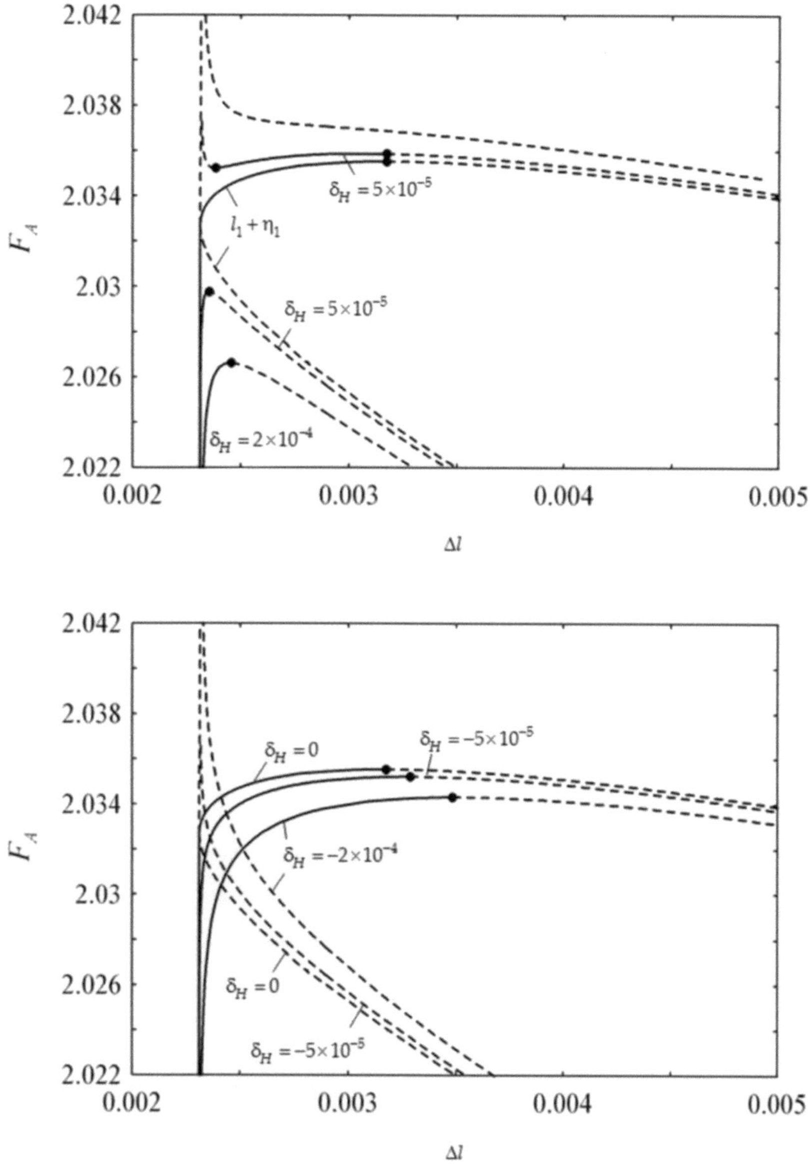

Figure 13. Load-deflection curves for (a) δ_H =0, 5×10^{-5}, and 2×10^{-4}; (b) δ_H =0, -5×10^{-5}, and -2×10^{-4}, respectively. The solid and dashed curves represent stable and unstable deformations, respectively.

whole lower branch is unstable, as in the inset of Figure 2. When δ_H increases by a small amount, the load-deflection curves degenerates into two branches veering away from each other. When $\delta_H = 5\times10^{-5}$, the stable range on the upper branch shrinks. When δ_H continues to increase to 2×10^{-4}, the stable range on the upper branch disappears altogether. For the lower branch, there exists a limit point. The locus with positive slope before the limit point is stable.

Figure 13(b) shows another scenario when δ_H varies from 0 to -5×10^{-5}, and -2×10^{-4}. It is observed that for a negative δ_H, the sharp corner degenerates into two smooth curves crossing each other. The one emerging from the lower part has a stable range which ends at the peak of the curve. The other curve emerging from the top is unstable all the way.

6. Conclusions

In this chapter we introduce a vibration method which is suitable to analyze the stability of a constrained elastica. A planar elastica constrained by a space-fixed point constraint is used to demonstrate the method. Generally speaking, static analysis allows one to find all the possible equilibrium configurations of a constrained elastica. In order to predict how the elastica behaves in reality, the stability of these equilibrium configurations needs to be determined. The key of the vibration method is to take into account the sliding between the elastica and the unilateral constraint during vibration. In order to accomplish this, Eulerian coordinates are defined to specify the positions of the material points on the elastica. After transforming the governing equations and the boundary conditions from the Lagrangian description to the Eulerian one, the natural frequencies and the vibration mode shapes of the constrained elastica can be calculated. The vibration method is applied to an elastica constrained by a point constraint in this chapter. The same principles can be extended to other similar problems as well, for instance; multiple point constraints (Chen et al., 2010) and plane constraints (Ro et al., 2010).

Author details

Jen-San Chen and Wei-Chia Ro
Department of Mechanical Engineering, National Taiwan University, Taipei, Taiwan

7. References

Adams, G.G., Benson R.C. (1986) Postbuckling of an Elastic Plate in a Rigid Channel. International Journal of Mechanical Sciences 28, 153–162.

Adan, N., Sheinman, I., Altus, E. (1994) Post-Buckling Behavior of Beams Under Contact Constraints. Journal of Applied Mechanics 61, 764–772.

Chai, H. (1990) Three-Dimensional Analysis of Thin-Film Debonding. International Journal of fracture 46, 237-256.

Chai, H. (1998) The Post-Buckling Behavior of a Bilaterally Constrained Column. Journal of the Mechanics and Physics of Solids 46, 1155–1181.

Chai, H. (2002) On the Post-Buckling Behavior of Bilaterally Constrained Plates. International Journal of Solids and Structures 39, 2911-2926.

Chateau, X., Nguyen, Q.S (1991) Buckling of Elastic Structures in Unilateral Contact With or Without Friction. European Journal of Mechanics, A/Solids 1, 71–89.

Chen, J.-S., Li C.-W. (2007) Planar Elastica Inside a Curved Tube with Clearance. International Journal of Solids and Structures 44, 6173-6186.

Chen, J.-S., Li, H.-C., Ro, W.-C. (2010) Slip-Through of a Heavy Elastica on Point Supports. International Journal of Solids and Structures, 47, 261–268.

Chen, J.-S., Lin, Y.-Z. (2008) Snapping of a Planar Elastica With Fixed End Slopes. ASME Journal of Applied Mechanics 75, 041024.

Chen, J.-S., Ro, W.-C. (2010) Deformations and Stability of an Elastica Subjected to an Off-Axis Point Constraint. ASME Journal of Applied Mechanics 77, 031006.

Denoel, V., Detournay, E. (2011) Eulerian Formulation of Constrained Elastica. International Journal of Solids and Structures 48, 625-636.

Domokos, G., Holmes, P., Royce, B. (1997) Constrained Euler Buckling. Journal of Nonlinear Science 7, 281-314.

Feodosyev, V.I. (1977) Selected Problems and Questions in Strength of Materials. Mir, Moscow.

Holmes, P., Domokos, G., Schmitt, J., Szeberenyi, I. (1999) Constrained Euler Buckling: an Interplay of Computation and Analysis. Computer Methods in Applied Mechanics and Engineering 170, 175-207.

Kuru, E., Martinez, A., Miska, S., Qiu, W. (2000) The Buckling Behavior of Pipes and its Influence on the Axial Force Transfer in Directional wells. ASME Journal of Energy Resources Technology 122, 129-135.

Lu, Z.-H., Chen, J.-S. (2008) Deformations of a Clamped-Clamped Elastica Inside a Circular Channel With Clearance. International Journal of Solids and Structures 45, 2470-2492.

Patricio, P., Adda-Bedia, M., Ben Amar, M. (1998) An Elastica Problem: Instabilities of an Elastic Arch. Physica D 124, 285-295.

Perkins, N. C. (1990) Planar Vibration of an Elastica Arch: Theory and Experiment. ASME Journal of Vibration and Acoustics 112, 374–379.

Ro, W.-C., Chen, J.-S., Hung, S.-Y. (2010) Vibration and Stability of a Constrained Elastica with Variable Length. International Journal of Solids and Structures, 47, 2143-2154.

Roman, B., Pocheau, A. (1999) Buckling Cascade of Thin Plates: Forms, Constraints and Similarity. Europhysics Letters 46, 602–608.

Roman, B., Pocheau, A. (2002) Postbuckling of Bilaterally Constrained Rectangular Thin Plates. Journal of the Mechanics and Physics of Solids 50, 2379-2401.

Santillan, S.T., Virgin, L.N., Plaut, R.H. (2006) Post-Buckling and Vibration of Heavy Beam on Horizontal or Inclined Rigid Foundation. Journal of Applied Mechanics 73, 664-671.

Vaillette, D.P., Adams, G.G. (1983) An Elastic Beam Contained in a Frictionless Channel. ASME Journal of Applied Mechanics 50, 693–694.

Chapter 4

Advanced Analysis of Space Steel Frames

Huu-Tai Thai

Additional information is available at the end of the chapter

http://dx.doi.org/10.5772/45808

1. Introduction

This chapter presents advanced analysis methods for space steel frames which consider both geometric and material nonlinearities. The geometric nonlinearities come from second-order $P-\Delta$ and $P-\delta$ effects (see Fig. 1.) as well as geometric imperfections, while the material nonlinearities are due to gradual yielding associated with residual stresses and flexure. The $P-\Delta$ effect results from the axial force acting through the relative displacement of the ends of the member, so it is referred to as a member chord rotation effect. The $P-\Delta$ effect is accounted in the second-order analysis by updating the configuration of the structure during the analysis process. The $P-\delta$ effect is caused by the axial force acting through the lateral displacement of the member relative to its chord, so it is referred to as a member curvature effect. The $P-\delta$ effect can be captured by using stability functions. Since the stability functions are derived from the closed-form solution of a beam-column subjected to end forces, they can accurately capture the $P-\delta$ effect by using only one element per member. Another way to capture the $P-\delta$ effect without using stability functions is to divide the member into many elements, and consequently, the $P-\delta$ effect is transformed to the $P-\Delta$ effect.

Geometric imperfections result from unavoidable errors during the fabrication or erection. There are three methods to model the geometric imperfections: (1) the explicit imperfection modeling, (2) the equivalent notional load, and (3) the further reduced tangent modulus. The explicit imperfection modeling for braced and unbraced members is illustrated in Fig. 2(a). For braced members, out-of-straightness is used instead of out-of-plumbness. This is due to the fact that the $P-\Delta$ effect due to the out-of-plumbness is vanished by braces. The limitation of this method is that it requires the determination of the direction of geometric imperfections which is often difficult in a large structural system. In the equivalent notional load method, the geometric imperfections are replaced by equivalent notional lateral loads in proportion to the gravity loads acting on the story as described in Fig. 2(b). The drawback of this method is that the gravity loads must be known in advance to determine the notional loads before analysis. Another way to account for the geometric imperfections is to further reduce the tangent modulus. The advantage of this method over the explicit imperfection modeling and equivalent notional load methods is its convenience and simplicity because it eliminates the inconvenience of explicit imperfection modeling and equivalent notional load methods.

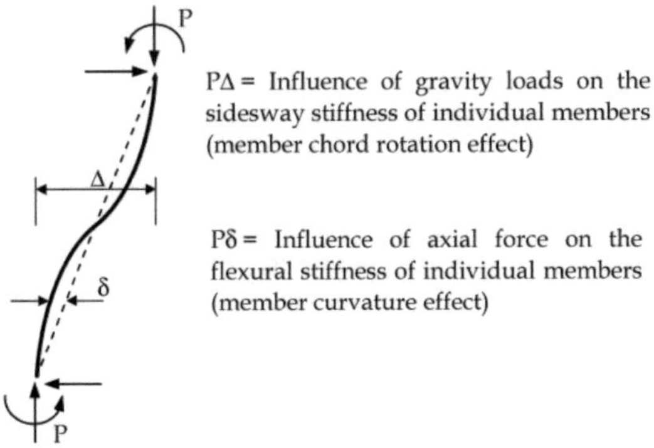

Figure 1. The $P-\delta$ and $P-\Delta$ effects of a beam-column

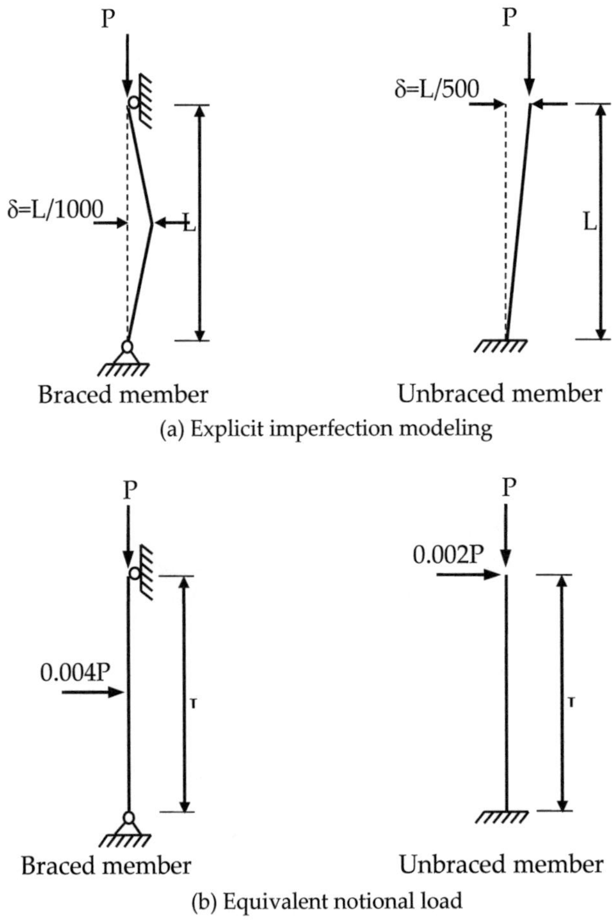

Figure 2. Geometric imperfection methods

Residual stresses are created in the hot-rolled sections due to uneven cooling of the cross-section. Typical residual stress pattern for a hot-rolled wide flange section is illustrated in Fig. 3. When a member is subjected to a compressive force, the fibers which have the highest values of compressive residual stress will yield first, and the fibers with the tensile stress will yield last. It means that the yielding over the cross-section is a gradual process. Hence, the stress-strain curve for a stub column is smooth instead of linear elastic-perfectly plastic in the case of coupon as shown in Fig. 4(a). The gradual yielding over the cross-section is caused not only by residual stress but also by flexure as shown in Fig. 4(b). Although the stress-strain relationship of steel is assumed to be linear elastic-perfectly plastic, the moment-curvature relationship has a smooth transition from elastic to fully plastic. This is because the section starts to yield gradually from extreme fibers which have the highest stresses. Material nonlinearities can be taken into account using various methods based on the degree of refinement used to represent yielding. The elastic plastic hinge method allows a drastic simplification, while the plastic zone method uses the greatest refinement.

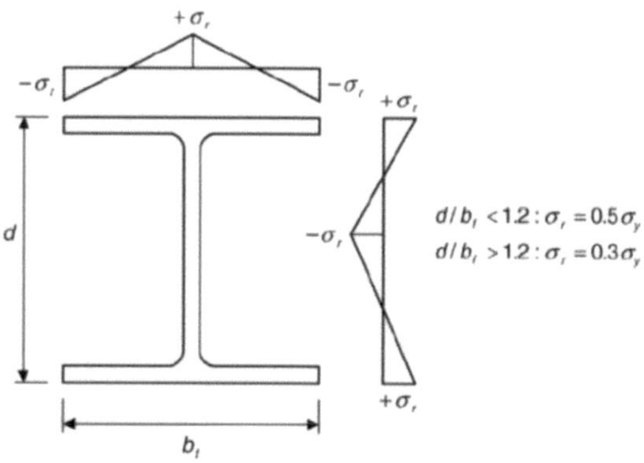

Figure 3. Typical residual stress pattern for a hot-rolled wide flange section

In the current design approach, the strength and stability of a structural system and its members are treated separately, and hence, the information about the failure modes of a structural system is not provided. This disadvantage is overcome by using a second-order inelastic analysis called "advanced analysis". Advanced analysis indicates any methods that efficiently and accurately capture the behavior and the strength of a structural system and its component members. This chapter will present two advanced analysis methods: (1) the refined plastic hinge method and (2) the fiber method. In these methods, the geometric nonlinearities are captured using the stability functions, while the material nonlinearities are considered using the refined plastic hinge model and fiber model. The benefit of employing the stability functions is that it can accurately capture geometrical nonlinear effects by using only one element per member, and hence, this leads to a high computational efficiency as demonstrated by the works of Thai and Kim (2008; 2009; 2011b; 2011c; 2011d; 2012).

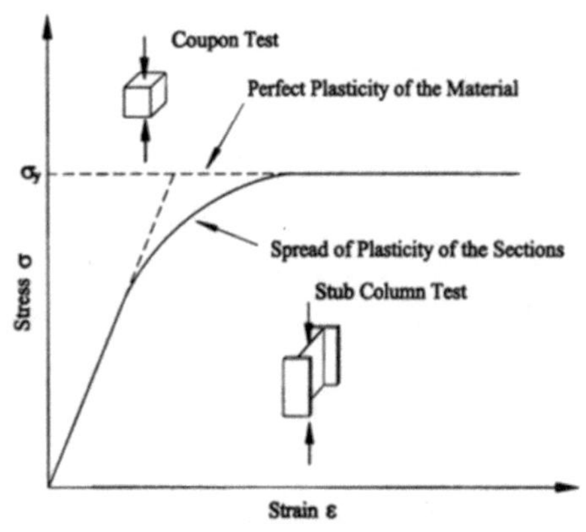

(a) Due to residual stress

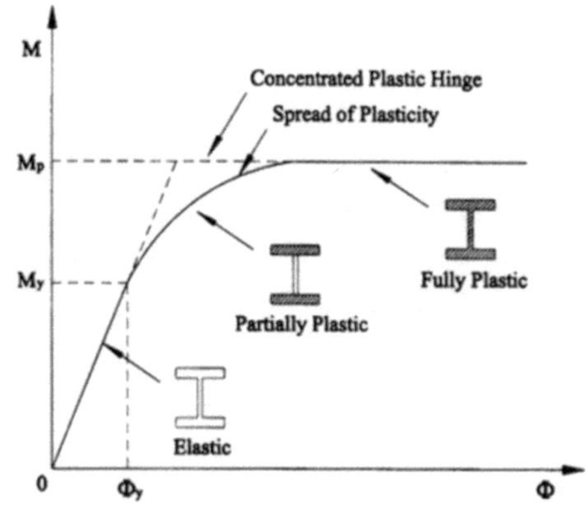

(b) Due to flexure

Figure 4. Gradual yielding of steel member

2. Advanced analysis

2.1. Stability functions accounting for second-order effects

Considering a beam-column element subjected to end moments and axial force as shown in Fig. 5. Using the free-body diagram of a segment of a beam-column element of length x, the external moment acting on the cut section is

$$M_{ext} = M_A + Py - \frac{M_A + M_B}{L} x = -EIy'' \tag{1}$$

where E, I, and L are the elastic modulus, moment of inertia, and length of an element, respectively.

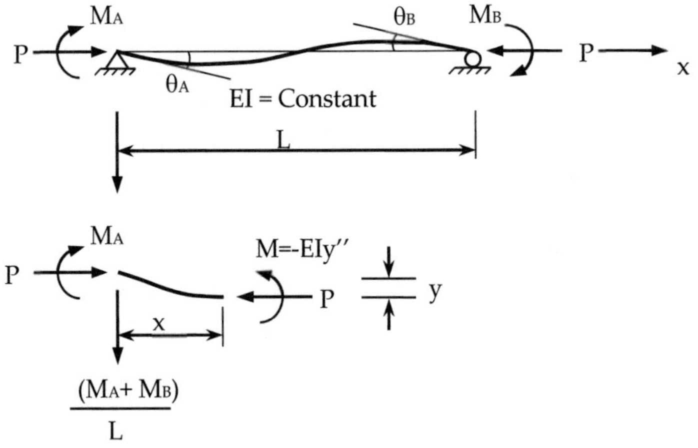

Figure 5. Beam-column with double-curvature bending

Using $k^2 = P / EI$, Eq. (1) is rewritten as

$$y'' + k^2 P = \frac{M_A + M_B}{EIL} x - \frac{M_A}{EI} \tag{2}$$

The general solution of Eq. (2) is

$$y = C_1 \sin kx + C_2 \cos kx + \frac{M_A + M_B}{EILk^2} x - \frac{M_A}{EIk^2} \tag{3}$$

The constants C_1 and C_2 are determined using the boundary conditions $y(0) = y(L) = 0$

$$C_1 = -\frac{M_A \cos kL + M_B}{EIk^2 \sin kL} \text{ and } C_2 = \frac{M_A}{EIk^2} \tag{4}$$

Substituting Eq. (4) into Eq. (3), the deflection y can be written as

$$y = -\frac{1}{EIk^2}\left[\frac{\cos kL}{\sin kL}\sin kx - \cos kx - \frac{x}{L} + 1\right]M_A - \frac{1}{EIk^2}\left[\frac{1}{\sin kL}\sin kx - \frac{x}{L}\right]M_B \tag{5}$$

and rotation y' is given as

$$y' = -\frac{1}{EIk}\left[\frac{\cos kL}{\sin kL}\cos kx + \sin kx - \frac{1}{kL}\right]M_A - \frac{1}{EIk}\left[\frac{1}{\sin kL}\cos kx - \frac{1}{kL}\right]M_B \tag{6}$$

The end rotation θ_A and θ_B can be obtained as

$$\theta_A = y'(0) = -\frac{1}{EIk}\left[\frac{\cos kL}{\sin kL} - \frac{1}{kL}\right]M_A - \frac{1}{EIk}\left[\frac{1}{\sin kL} - \frac{1}{kL}\right]M_B \quad \text{(a)}$$
$$\theta_B = y'(L) = -\frac{1}{EIk}\left[\frac{1}{\sin kL} - \frac{1}{kL}\right]M_A - \frac{1}{EIk}\left[\frac{\cos kL}{\sin kL} - \frac{1}{kL}\right]M_B \quad \text{(b)} \tag{7}$$

Eq. (7) can be written in matrix from as

$$\begin{Bmatrix} M_A \\ M_B \end{Bmatrix} = \frac{EI}{L}\begin{bmatrix} S_1 & S_2 \\ S_2 & S_1 \end{bmatrix}\begin{Bmatrix} \theta_A \\ \theta_B \end{Bmatrix} \tag{8}$$

where S_1 and S_2 are the stability functions defined as

$$S_1 = \frac{kL(\sin kL - kL\cos kL)}{2 - 2\cos kL - kL\sin kL} \quad \text{(a)}$$
$$S_2 = \frac{kL(kL - \sin kL)}{2 - 2\cos kL - kL\sin kL} \quad \text{(b)} \tag{9}$$

S_1 and S_2 account for the coupling effect between axial force and bending moments of the beam-column member. For members subjected to an axial force that is tensile rather than compressive, the stability functions are redefined as

$$S_1 = \frac{kL\left(kL\cosh kL - \sinh kL\right)}{2 - 2\cosh kL - kL\sinh kL} \quad \text{(a)}$$
$$S_2 = \frac{kL\left(\sinh kL - kL\right)}{2 - 2\cosh kL - kL\sinh kL} \quad \text{(b)} \tag{10}$$

Eqs. (9) and (10) are indeterminate when the axial force is zero (i.e. $kL = 0$). To overcome this problem, the following simplified equations are used to approximate the stability functions when the axial force in the member falls within the range of -2.0 ≤ ρ ≤ 2.0

$$S_1 = 4 + \frac{2\pi^2 \rho}{15} - \frac{(0.01\rho + 0.543)\rho^2}{4 + \rho} - \frac{(0.004\rho + 0.285)\rho^2}{8.183 + \rho} \quad \text{(a)}$$
$$S_2 = 2 - \frac{\pi^2 \rho}{30} + \frac{(0.01\rho + 0.543)\rho^2}{4 + \rho} - \frac{(0.004\rho + 0.285)\rho^2}{8.183 + \rho} \quad \text{(b)} \tag{11}$$

where $\rho = P / P_e = P/(\pi^2 EI/L^2) = (kL / \pi)^2$. For most practical applications, it gives excellent correlation to the "exact" expressions given by Eqs. (9) and (10). However, for ρ other than the range of -2.0 ≤ ρ ≤ 2.0, the conventional stability functions in Eqs. (9) and (10) should be used. The incremental member force and deformation relationship of a three-dimensional beam-column element under axial force and end moments can be written as

$$\begin{Bmatrix} \Delta P \\ \Delta M_{yA} \\ \Delta M_{yB} \\ \Delta M_{zA} \\ \Delta M_{zB} \\ \Delta T \end{Bmatrix} = \begin{bmatrix} \frac{EA}{L} & 0 & 0 & 0 & 0 & 0 \\ 0 & S_{1y}\frac{EI_y}{L} & S_{2y}\frac{EI_y}{L} & 0 & 0 & 0 \\ 0 & S_{2y}\frac{EI_y}{L} & S_{1y}\frac{EI_y}{L} & 0 & 0 & 0 \\ 0 & 0 & 0 & S_{1z}\frac{EI_z}{L} & S_{2z}\frac{EI_z}{L} & 0 \\ 0 & 0 & 0 & S_{2z}\frac{EI_z}{L} & S_{1z}\frac{EI_z}{L} & 0 \\ 0 & 0 & 0 & 0 & 0 & \frac{GJ}{L} \end{bmatrix} \begin{Bmatrix} \Delta \delta \\ \Delta \theta_{yA} \\ \Delta \theta_{yB} \\ \Delta \theta_{zA} \\ \Delta \theta_{zB} \\ \Delta \phi \end{Bmatrix} \tag{12}$$

where ΔP, ΔM_{yA}, ΔM_{yB}, ΔM_{zA}, ΔM_{zB}, and ΔT are the incremental axial force, end moments with respect to y and z axes, and torsion, respectively; $\Delta\delta$, $\Delta\theta_{yA}$, $\Delta\theta_{yB}$, $\Delta\theta_{zA}$, $\Delta\theta_{zB}$, and $\Delta\phi$ are the incremental axial displacement, the end rotations, and the angle of twist, respectively; S_{1n} and S_{2n} are stability functions with respect to n axis $(n = y,z)$ given in Eqs. (9) and (10); and EA, EI_n, and GJ denote the axial, bending, and torsional stiffness, respectively.

2.2. Refined plastic hinge model accounting for inelastic effects

The refined plastic hinge model is an improvement of the elastic plastic hinge one. Two modifications are made to account for a smooth degradation of plastic hinge stiffness: (1) the tangent modulus concept is used to capture the residual stress effect along the length of the member, and (2) the parabolic function is adopted to represent the gradual yielding effect in forming plastic hinges. The inelastic behavior of the member is modeled in terms of member force instead of the detailed level of stresses and strains as used in the plastic zone method. As a result, the refined plastic hinge method retains the simplicity of the elastic plastic hinge method, but it is sufficiently accurate for predicting the strength and stability of a structural system and its component members.

2.2.1. Gradual yielding due to residual stresses

The Column Research Council (CRC) tangent modulus concept is employed to account for the gradual yielding along the member length due to residual stresses. The elastic modulus E (instead of moment of inertia I) is reduced to account for the reduction of the elastic portion of the cross-section since the reduction of the elastic modulus is easier to implement than a new moment of inertia for every different section. The rate of reduction in stiffness is different in the weak and strong direction, but this is not considered since the dramatic degradation of weak-axis stiffness is compensated for by the substantial weak-axis plastic strength. This simplification makes the present method more practical. The CRC tangent modulus E_t can be written as

$$\begin{aligned} &E_t = 1.0E \quad \text{for} \quad P \le 0.5P_y && \text{(a)} \\ &E_t = 4\frac{P}{P_y}\left(1 - \frac{P}{P_y}\right)E \quad \text{for} \quad 0.5P_y < P \le P_y && \text{(b)} \\ &E_t = 0 \quad \text{for} \quad P > P_y && \text{(c)} \end{aligned} \tag{13}$$

Equation (13) is plotted in Fig. 6. The tangent modulus E_t is reduced from the elastic value when $P > 0.5P_y$.

2.2.2. Gradual yielding due to flexure

The tangent modulus concept is suitable for the member subjected to axial force, but not adequate for cases of both axial force and bending moment. A gradual stiffness degradation model for a plastic hinge is required to represent the partial plastification effects associated with flexure. The parabolic function is used to represent the smooth transition from elastic stiffness at the onset of yielding to the stiffness associated with a full plastic hinge. The parabolic function η representing the gradual stiffness degradation is obtained based on a calibration with plastic zone solutions of simple portal frames and beam-columns. It should be noted that only a simple relationship for η is required to describe the degradation in stiffness associated with flexure. Although more complicated expressions for η can be proposed, simple expression for η is needed for keeping the analysis model simple and straightforward.

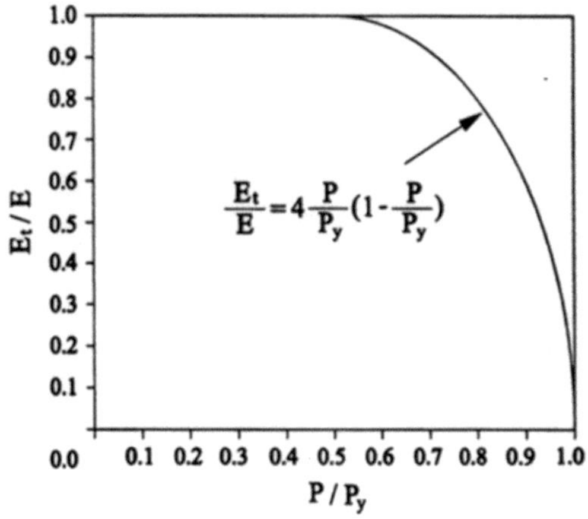

Figure 6. Stiffness reduction due to residual stress

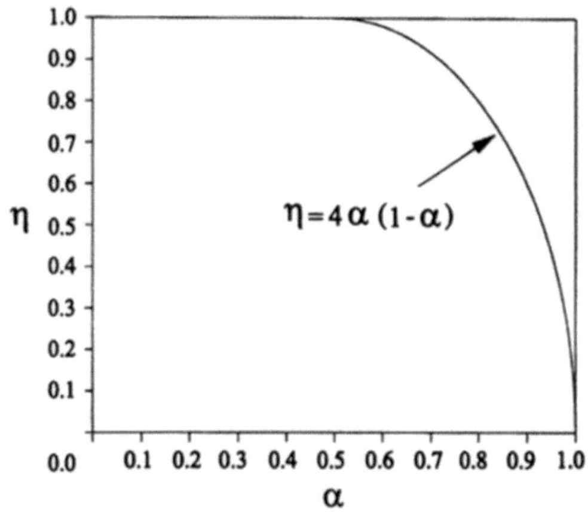

Figure 7. Stiffness degradation function

The value of parabolic function η is equal to 1.0 when the element is elastic, and zero when a plastic hinge is formed. The parabolic function η can be expressed as (see Fig. 7.)

$$
\begin{aligned}
&\eta = 1.0 \ \text{for} \ \alpha \le 0.5 && \text{(a)} \\
&\eta = 4\alpha(1-\alpha) \ \text{for} \ 0.5 < \alpha \le 1.0 && \text{(b)} \\
&\eta = 0 \ \text{for} \ \alpha > 1 && \text{(c)}
\end{aligned}
\qquad (14)
$$

where α is the force-state parameter which can be expressed by AISC-LRFD or modified Orbison yield surfaces as (seeFig. 8.).

For AISC-LRFD yield surface (AISC, 2005)

$$\alpha = p + \frac{8}{9}m_y + \frac{8}{9}m_z \text{ for } p \geq \frac{2}{9}m_y + \frac{2}{9}m_z \quad \text{(a)}$$

$$\alpha = \frac{p}{2} + m_y + m_z \text{ for } p < \frac{2}{9}m_y + \frac{2}{9}m_z \quad \text{(b)} \qquad (15)$$

For modified Orbison yield surface (McGuire et al., 2000)

$$\alpha = p^2 + m_z^2 + m_y^4 + 3.5p^2m_z^2 + 3.0p^6m_y^2 + 4.5m_z^4m_y^2 \qquad (16)$$

where $p = P / P_y$, $m_z = M_z / M_{pz}$ (strong-axis), $m_y = M_y / M_{py}$ (weak-axis); P_y, M_{yp}, M_{zp} are axial load, and plastic moment capacity of the cross-section about y and z axes.

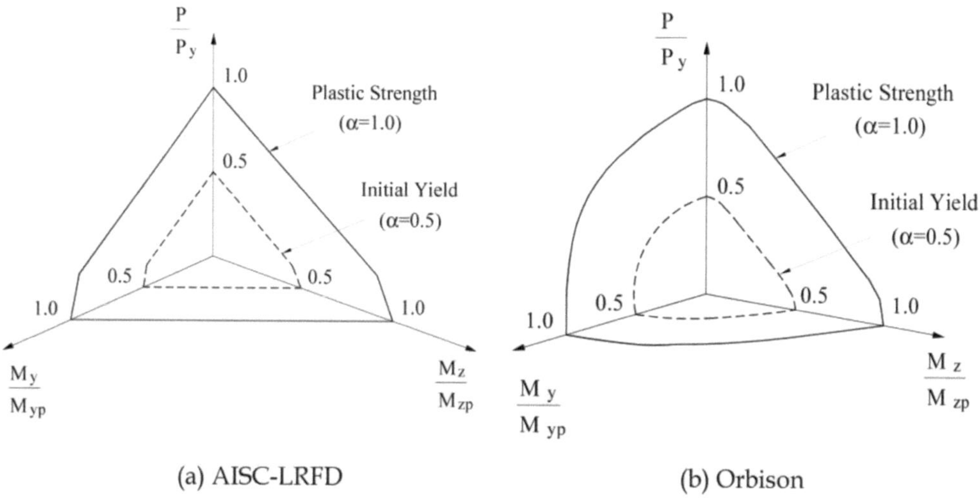

Figure 8. Plastification surface

When the force point moves inside or along the initial yield surface $(\alpha \leq 0.5)$, the element remains fully elastic (i.e. no stiffness reduction, $\eta = 1.0$). If the force point moves beyond the initial yield surface and inside the full yield surface $(0.5 < \alpha \leq 1.0)$, the element stiffness is reduced to account for the effect of plastification at the element end. The reduction of element stiffness is assumed to vary according to the parabolic function in the Eq. (15b). When member forces violate the plastic strength surface $(\alpha > 1.0)$, the member forces will be scaled down to move the force point return the yield surface based on incremental-iterative scheme.

When the parabolic function for a gradual yielding is active at both ends of an element, the incremental member force and deformation relationship in Eq. (12) is modified as

$$\begin{Bmatrix} \Delta P \\ \Delta M_{yA} \\ \Delta M_{yB} \\ \Delta M_{zA} \\ \Delta M_{zB} \\ \Delta T \end{Bmatrix} = \begin{bmatrix} \frac{E_t A}{L} & 0 & 0 & 0 & 0 & 0 \\ 0 & k_{iiy} & k_{ijy} & 0 & 0 & 0 \\ 0 & k_{ijy} & k_{jjy} & 0 & 0 & 0 \\ 0 & 0 & 0 & k_{iiz} & k_{ijz} & 0 \\ 0 & 0 & 0 & k_{ijz} & k_{jjz} & 0 \\ 0 & 0 & 0 & 0 & 0 & \frac{GJ}{L} \end{bmatrix} \begin{Bmatrix} \Delta\delta \\ \Delta\theta_{yA} \\ \Delta\theta_{yB} \\ \Delta\theta_{zA} \\ \Delta\theta_{zB} \\ \Delta\phi \end{Bmatrix} \tag{17}$$

where

$$\begin{aligned} k_{iiy} &= \eta_A (S_1 - \frac{S_2^2}{S_1}(1-\eta_B))\frac{E_t I_y}{L} && \text{(a)} \\ k_{ijy} &= \eta_A \eta_B S_2 \frac{E_t I_y}{L} && \text{(b)} \\ k_{jjy} &= \eta_B (S_1 - \frac{S_2^2}{S_1}(1-\eta_A))\frac{E_t I_y}{L} && \text{(c)} \\ k_{iiz} &= \eta_A (S_3 - \frac{S_4^2}{S_3}(1-\eta_B))\frac{E_t I_z}{L} && \text{(d)} \\ k_{ijz} &= \eta_A \eta_B S_4 \frac{E_t I_z}{L} && \text{(e)} \\ k_{jjz} &= \eta_B (S_3 - \frac{S_4^2}{S_3}(1-\eta_A))\frac{E_t I_z}{L} && \text{(f)} \end{aligned} \tag{18}$$

where η_A and η_B are the values of parabolic functions at the ends A and B, respectively.

2.3. Fiber model accounting for inelastic effects

The concept of fiber model is presented in Fig. 9. In this model, the element is divided into a number of monitored sections represented by the integration points. Each section is further divided into m fibers and each fiber is represented by its area A_i and coordinate location corresponding to its centroid (y_i, z_i). The inelastic effects are captured by tracing the uniaxial stress-strain relationship of each fiber on the cross sections located at the selected integration points along the member length.

The incremental force and deformation relationship, Eq. (12), which accounts for the $P-\delta$ effect can be rewritten in symbolic form as

$$\{\Delta F\} = [K_e]\{\Delta d\} \tag{19}$$

where

$$\{\Delta F\} = \begin{bmatrix} \Delta P & \Delta M_{yA} & \Delta M_{yB} & \Delta M_{zA} & \Delta M_{zB} & \Delta T \end{bmatrix}^T \tag{20}$$

$$\{\Delta d\} = \begin{bmatrix} \Delta \delta & \Delta \theta_{yA} & \Delta \theta_{yB} & \Delta \theta_{zA} & \Delta \theta_{zB} & \Delta \phi \end{bmatrix}^T \tag{21}$$

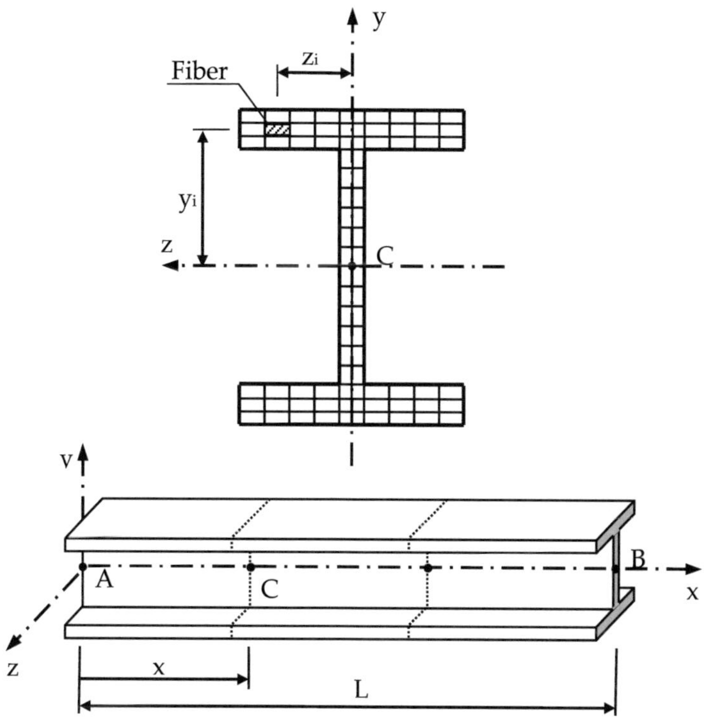

Figure 9. Fiber hinge model

$$K_e = \begin{bmatrix} \frac{EA}{L} & 0 & 0 & 0 & 0 & 0 \\ 0 & S_{1y}\frac{EI_y}{L} & S_{2y}\frac{EI_y}{L} & 0 & 0 & 0 \\ 0 & S_{2y}\frac{EI_y}{L} & S_{1y}\frac{EI_y}{L} & 0 & 0 & 0 \\ 0 & 0 & 0 & S_{1z}\frac{EI_z}{L} & S_{2z}\frac{EI_z}{L} & 0 \\ 0 & 0 & 0 & S_{2z}\frac{EI_z}{L} & S_{1z}\frac{EI_z}{L} & 0 \\ 0 & 0 & 0 & 0 & 0 & \frac{GJ}{L} \end{bmatrix} \tag{22}$$

in which the axial stiffness EA, bending stiffness EI_n, and torsional stiffness GJ of the fiber element can be obtained as

$$\begin{aligned}
EA &= \sum_{j=1}^{h} w_j \left(\sum_{i=1}^{m} E_i A_i \right)_j && \text{(a)} \\
EI_y &= \sum_{j=1}^{h} w_j \left(\sum_{i=1}^{m} E_i A_i z_i^2 \right)_j && \text{(b)} \\
EI_z &= \sum_{j=1}^{h} w_j \left(\sum_{i=1}^{m} E_i A_i y_i^2 \right)_j && \text{(c)} \\
GJ &= \sum_{j=1}^{h} w_j \left[\sum_{i=1}^{m} \left(y_i^2 + z_i^2 \right) G_i A_i \right]_j && \text{(d)}
\end{aligned} \tag{23}$$

in which h is the total number of monitored sections along an element; m is the total number of fiber divided on the monitored cross-section; w_j is the weighting factor of the jth section; E_i and G_i are the tangent and shear modulus of ith fiber, respectively; y_i and z_i are the coordinates of ith fiber in the cross-section. The element stiffness matrix is evaluated numerically by the Gauss-Lobatto integration scheme since this method allows for two integration points to coincide with the end sections of the elements. Since inelastic behavior in beam elements often concentrates at the end of member, the monitoring of the end sections of the element is advantageous from the standpoint of accuracy and numerical stability. By contrast, the outermost integration points of the classical Gauss integration method only approach the end sections with increasing order of integration, but never coincide with the end sections and, hence, result in overestimation of the member strength (Spacone et al., 1996).

Section deformations are represented by three strain resultants: the axial strain ε along the longitudinal axis and two curvatures χ_z and χ_y with respect to z and y axes, respectively. The corresponding force resultants are the axial force N and two bending moments M_z and M_y. The section forces and deformations are grouped in the following vectors:

$$\text{Section force vector } \{Q\} = \begin{bmatrix} M_z & M_y & N \end{bmatrix}^T \tag{24}$$

$$\text{Section deformation vector } \{q\} = \begin{bmatrix} \chi_z & \chi_y & \varepsilon \end{bmatrix}^T \tag{25}$$

The incremental section force vector at each integration points is determined based on the incremental element force vector $\{\Delta F\}$ as

$$\{\Delta Q\} = \left[B(x) \right] \{\Delta F\} \tag{26}$$

where $\left[B(x)\right]$ is the force interpolation function matrix given as

$$\left[B(x)\right]=\begin{bmatrix}0 & 0 & 0 & (x/L-1) & x/L & 0\\ 0 & (x/L-1) & x/L & 0 & 0 & 0\\ 1 & 0 & 0 & 0 & 0 & 0\end{bmatrix} \tag{27}$$

The section deformation vector is determined based on the section force vector as

$$\{\Delta q\}=\left[k_{\text{sec}}\right]^{-1}\{\Delta Q\} \tag{28}$$

where $\left[k_{\text{sec}}\right]$ is the section stiffness matrix given as

$$\left[k_{\text{sec}}\right]=\begin{bmatrix}\sum_{i=1}^{m}E_iA_iy_i^2 & \sum_{i=1}^{m}E_iA_iy_iz_i & \sum_{i=1}^{m}E_iA_i(-y_i)\\ \sum_{i=1}^{m}E_iA_iy_iz_i & \sum_{i=1}^{m}E_iA_iz_i^2 & \sum_{i=1}^{m}E_iA_iz_i\\ \sum_{i=1}^{m}E_iA_i(-y_i) & \sum_{i=1}^{m}E_iA_iz_i & \sum_{i=1}^{m}E_iA_i\end{bmatrix} \tag{29}$$

Following the hypothesis that plane sections remain plane and normal to the longitudinal axis, the incremental uniaxial fiber strain vector is computed based on the incremental section deformation vector as

$$\{\Delta e\}=\left[\Gamma\right]\{\Delta q\} \tag{30}$$

where $\left[\Gamma\right]$ is the linear geometric matrix given as follows

$$\left[\Gamma\right]=\begin{bmatrix}-y_1 & z_1 & 1\\ -y_2 & z_2 & 1\\ \cdots & \cdots & \cdots\\ -y_m & z_m & 1\end{bmatrix} \tag{31}$$

Once the incremental fiber strain is evaluated, the incremental fiber stress is computed based on the stress-strain relationship of material model. The tangent modulus of each fiber is updated from the incremental fiber stress and incremental fiber strain as

$$E_i=\frac{\Delta\sigma_i}{\Delta e_i} \tag{32}$$

Eq. (32) leads to updating of the element stiffness matrix $\left[K_e\right]$ in Eq. (22) and section stiffness matrix $\left[k_{\text{sec}}\right]$ in Eq. (29) during the iteration process. Based on the new tangent modulus of Eq. (32), the location of the section centroid is also updated during the incremental load steps to take into account the distribution of section plasticity. The section resisting forces are computed by summation of the axial force and biaxial bending moment contributions of all fibers as

$$\{Q_R\} = \begin{Bmatrix} M_z \\ M_y \\ N \end{Bmatrix} = \begin{Bmatrix} \sum_{i=1}^{m} \sigma_i A_i (-y_i) \\ \sum_{i=1}^{m} \sigma_i A_i z_i \\ \sum_{i=1}^{m} \sigma_i A_i \end{Bmatrix} \tag{33}$$

2.4. Shear deformation effect

To account for transverse shear deformation effect in a beam-column element, the member force and deformation relationship of beam-column element in Eq. (12) should be modified. The flexibility matrix can be obtained by inversing the flexural stiffness matrix as

$$\begin{Bmatrix} \Delta\theta_{MA} \\ \Delta\theta_{MB} \end{Bmatrix} = \begin{bmatrix} \dfrac{k_{jj}}{k_{ii}k_{jj} - k_{ij}^2} & \dfrac{-k_{ij}}{k_{ii}k_{jj} - k_{ij}^2} \\ \dfrac{-k_{ij}}{k_{ii}k_{jj} - k_{ij}^2} & \dfrac{k_{ii}}{k_{ii}k_{jj} - k_{ij}^2} \end{bmatrix} \begin{Bmatrix} \Delta M_A \\ \Delta M_B \end{Bmatrix} \tag{34}$$

where $\Delta\theta_{MA}$ and $\Delta\theta_{MB}$ are the slope of the neutral axis due to bending moment. The flexibility matrix corresponding to shear deformation can be written as

$$\begin{Bmatrix} \Delta\theta_{SA} \\ \Delta\theta_{SB} \end{Bmatrix} = \begin{bmatrix} \dfrac{1}{GA_S L} & \dfrac{1}{GA_S L} \\ \dfrac{1}{GA_S L} & \dfrac{1}{GA_S L} \end{bmatrix} \begin{Bmatrix} \Delta M_A \\ \Delta M_B \end{Bmatrix} \tag{35}$$

where GA_S and L are shear stiffness and length of the element, respectively. The total rotations at the two ends A and B are obtained by combining Eqs. (34) and (35) as

$$\begin{Bmatrix} \Delta\theta_A \\ \Delta\theta_B \end{Bmatrix} = \begin{Bmatrix} \Delta\theta_{MA} \\ \Delta\theta_{MB} \end{Bmatrix} + \begin{Bmatrix} \Delta\theta_{SA} \\ \Delta\theta_{SB} \end{Bmatrix} \tag{36}$$

The basic force and deformation relationship including shear deformation is derived by inverting the flexibility matrix as

$$\begin{Bmatrix} \Delta M_A \\ \Delta M_B \end{Bmatrix} = \begin{bmatrix} \dfrac{k_{ii}k_{jj} - k_{ij}^2 + k_{ii}A_sGL}{k_{ii} + k_{jj} + 2k_{ij} + A_sGL} & \dfrac{-k_{ii}k_{jj} + k_{ij}^2 + k_{ij}A_sGL}{k_{ii} + k_{jj} + 2k_{ij} + A_sGL} \\ \dfrac{-k_{ii}k_{jj} + k_{ij}^2 + k_{ij}A_sGL}{k_{ii} + k_{jj} + 2k_{ij} + A_sGL} & \dfrac{k_{ii}k_{jj} - k_{ij}^2 + k_{jj}A_sGL}{k_{ii} + k_{jj} + 2k_{ij} + A_sGL} \end{bmatrix} \begin{Bmatrix} \Delta\theta_A \\ \Delta\theta_B \end{Bmatrix} \tag{37}$$

The member force and deformation relationship can be extended for three-dimensional beam-column element as

$$\begin{Bmatrix} \Delta P \\ \Delta M_{yA} \\ \Delta M_{yB} \\ \Delta M_{zA} \\ \Delta M_{zB} \\ \Delta T \end{Bmatrix} = \begin{bmatrix} \dfrac{EA}{L} & 0 & 0 & 0 & 0 & 0 \\ 0 & C_{iiy} & C_{ijy} & 0 & 0 & 0 \\ 0 & C_{ijy} & C_{jjy} & 0 & 0 & 0 \\ 0 & 0 & 0 & C_{iiz} & C_{ijz} & 0 \\ 0 & 0 & 0 & C_{ijz} & C_{jjz} & 0 \\ 0 & 0 & 0 & 0 & 0 & \dfrac{GJ}{L} \end{bmatrix} \begin{Bmatrix} \Delta\delta \\ \Delta\theta_{yA} \\ \Delta\theta_{yB} \\ \Delta\theta_{zA} \\ \Delta\theta_{zB} \\ \Delta\phi \end{Bmatrix} \tag{38}$$

in which

$$C_{iiy} = \frac{k_{iiy}k_{jjy} - k_{ijy}^2 + k_{iiy}A_{sz}GL}{k_{iiy} + k_{jjy} + 2k_{ijy} + A_{sz}GL} \quad \text{(a)}$$

$$C_{ijy} = \frac{-k_{iiy}k_{jjy} + k_{ijy}^2 + k_{ijy}A_{sz}GL}{k_{iiy} + k_{jjy} + 2k_{ijy} + A_{sz}GL} \quad \text{(b)}$$

$$C_{jjy} = \frac{k_{iiy}k_{jjy} - k_{ijy}^2 + k_{jjy}A_{sz}GL}{k_{iiy} + k_{jjy} + 2k_{ijy} + A_{sz}GL} \quad \text{(c)}$$

$$C_{iiz} = \frac{k_{iiz}k_{jjz} - k_{ijz}^2 + k_{iiz}A_{sy}GL}{k_{iiz} + k_{jjz} + 2k_{ijz} + A_{sy}GL} \quad \text{(d)} \qquad (39)$$

$$C_{ijz} = \frac{-k_{iiz}k_{jjz} + k_{ijz}^2 + k_{ijz}A_{sy}GL}{k_{iiz} + k_{jjz} + 2k_{ijz} + A_{sy}GL} \quad \text{(e)}$$

$$C_{jjz} = \frac{k_{iiz}k_{jjz} - k_{ijz}^2 + k_{jjz}A_{sy}GL}{k_{iiz} + k_{jjz} + 2k_{ijz} + A_{sy}GL} \quad \text{(f)}$$

where A_{sy} and A_{sz} are the shear areas with respect to y and z axes, respectively.

2.5. Element stiffness matrix

The incremental end forces and displacements used in Eq. (38) are shown in Fig. 10(a). The sign convention for the positive directions of element end forces and displacements of a frame member is shown in Fig. 10(b). By comparing the two figures, the equilibrium and kinematic relationships can be expressed in symbolic form as

$$\begin{aligned} \{f_n\} &= [T]_{6\times 12}^{T}\{F\} && \text{(a)} \\ \{d\} &= [T]_{6\times 12}\{d_L\} && \text{(b)} \end{aligned} \tag{40}$$

where $\{f_n\}$ and $\{d_L\}$ are the nodal force and nodal displacement vectors of the element expressed as

$$\begin{aligned} \{f_n\}^T &= \{r_{n1} \quad r_{n2} \quad r_3 \quad r_4 \quad r_5 \quad r_6 \quad r_7 \quad r_8 \quad r_9 \quad r_{10} \quad r_{11} \quad r_{12}\} && \text{(a)} \\ \{d_L\}^T &= \{d_1 \quad d_2 \quad d_3 \quad d_4 \quad d_5 \quad d_6 \quad d_7 \quad d_8 \quad d_9 \quad d_{10} \quad d_{11} \quad d_{12}\} && \text{(b)} \end{aligned} \tag{41}$$

and $\{F\}$ and $\{d\}$ are the basic member force and displacement vectors given in Eqs. (20) and (21), respectively. $[T]_{6\times 12}$ is a transformation matrix written as

$$[T]_{6\times 12} = \begin{bmatrix} -1 & 0 & 0 & 0 & 0 & 0 & 1 & 0 & 0 & 0 & 0 & 0 \\ 0 & 0 & -1/L & 0 & 1 & 0 & 0 & 0 & 1/L & 0 & 0 & 0 \\ 0 & 0 & -1/L & 0 & 0 & 0 & 0 & 0 & 1/L & 0 & 1 & 0 \\ 0 & 1/L & 0 & 0 & 0 & 1 & 0 & -1/L & 0 & 0 & 0 & 0 \\ 0 & 1/L & 0 & 0 & 0 & 0 & 0 & -1/L & 0 & 0 & 0 & 1 \\ 0 & 0 & 0 & 1 & 0 & 0 & 0 & 0 & 0 & -1 & 0 & 0 \end{bmatrix} \tag{42}$$

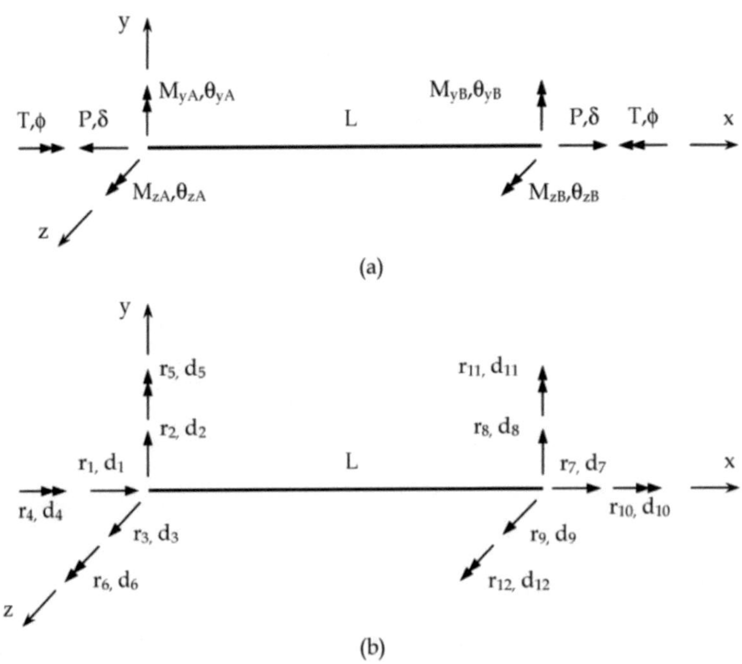

Figure 10. Force and displacement notations

Using the transformation matrix, the nodal force and nodal displacement relationship of element may be written as

$$\{f_n\} = [K_n]\{d_L\} \tag{43}$$

where $[K_n]$ is the element stiffness matrix expressed as

$$[K_n]_{12\times12} = [T]^T_{6\times12}[K_e]_{6\times6}[T]_{6\times12} \tag{44}$$

It should be noted that Eq. (43) is used for the beam-column member in which side-sway is restricted. If the beam-column member is permitted to sway, additional axial and shear forces will be induced in the member. These additional axial and shear forces due to member sway to the member end displacements can be related as

$$\{f_s\} = [K_s]\{d_L\} \tag{45}$$

where $[K_s]$ is the element stiffness matrix due to member sway expressed as

$$[K_s]_{12\times12} = \begin{bmatrix} [G_s] & -[G_s] \\ -[G_s]^T & [G_s] \end{bmatrix} \tag{46}$$

in which

$$[G_s] = \begin{bmatrix} 0 & (M_{zA}+M_{zB})/L^2 & (M_{yA}+M_{yB})/L^2 & 0 & 0 & 0 \\ (M_{zA}+M_{zB})/L^2 & P/L & 0 & 0 & 0 & 0 \\ (M_{yA}+M_{yB})/L^2 & 0 & P/L & 0 & 0 & 0 \\ 0 & 0 & 0 & 0 & 0 & 0 \\ 0 & 0 & 0 & 0 & 0 & 0 \\ 0 & 0 & 0 & 0 & 0 & 0 \end{bmatrix} \tag{47}$$

By combining Eqs. (43) and (47), the general force-displacement relationship of beam-column element obtained as

$$\{f_L\} = [K]\{d_L\} \tag{48}$$

where

$$\{f_L\} = \{f_n\} + \{f_s\} \tag{49}$$

$$[K] = [K_n] + [K_s] \tag{50}$$

2.6. Solution algorithm

The generalized displacement control method proposed by Yang and Shieh (1990) appears to be one of the most robust and effective method because of its general numerical stability and efficiency. This method is adopted herein to solve the nonlinear equilibrium equations. The incremental form of the equilibrium equation can be rewritten for the j th iteration of the i th incremental step as

$$\left[K_{j-1}^{i}\right]\left\{\Delta D_{j}^{i}\right\} = \lambda_{j}^{i}\left\{\hat{P}\right\} + \left\{R_{j-1}^{i}\right\} \tag{51}$$

where $\left[K_{j-1}^{i}\right]$ is the tangent stiffness matrix, $\left\{\Delta D_{j}^{i}\right\}$ is the displacement increment vector, $\left\{\hat{P}\right\}$ is the reference load vector, $\left\{R_{j-1}^{i}\right\}$ is the unbalanced force vector, and λ_{j}^{i} is the load increment parameter. According to Batoz and Dhatt (1979), Eq. (51) can be decomposed into the following equations:

$$\left[K_{j-1}^{i}\right]\left\{\Delta \hat{D}_{j}^{i}\right\} = \left\{\hat{P}\right\} \tag{52}$$

$$\left[K_{j-1}^{i}\right]\left\{\Delta \bar{D}_{j}^{i}\right\} = \left\{R_{j-1}^{i}\right\} \tag{53}$$

$$\left\{\Delta D_{j}^{i}\right\} = \lambda_{j}^{i}\left\{\Delta \hat{D}_{j}^{i}\right\} + \left\{\Delta \bar{D}_{j}^{i}\right\} \tag{54}$$

Once the displacement increment vector $\left\{\Delta D_{j}^{i}\right\}$ is determined, the total displacement vector $\left\{D_{j}^{i}\right\}$ of the structure at the end of j th iteration can be accumulated as

$$\left\{D_{j}^{i}\right\} = \left\{D_{j-1}^{i}\right\} + \left\{\Delta D_{j}^{i}\right\} \tag{55}$$

The total applied load vector $\left\{P_{j}^{i}\right\}$ at the j th iteration of the i th incremental step relates to the reference load vector $\left\{\hat{P}\right\}$ as

$$\left\{P_{j}^{i}\right\} = \Lambda_{j}^{i}\left\{\hat{P}\right\} \tag{56}$$

where the load factor Λ_{j}^{i} can be related to the load increment parameter λ_{j}^{i} by

$$\Lambda_{j}^{i} = \Lambda_{j-1}^{i} + \lambda_{j}^{i} \tag{57}$$

The load increment parameter λ_{j}^{i} is an unknown. It is determined from a constraint condition. For the first iterative step $(j=1)$, the load increment parameter λ_{j}^{i} is determined based on the generalized stiffness parameter (GSP) as

$$\lambda_{1}^{i} = \lambda_{1}^{1}\sqrt{|GSP|} \tag{58}$$

where λ_{1}^{1} is an initial value of load increment parameter, and the GSP is defined as

$$GSP = \frac{\left\{\Delta\hat{D}_1^1\right\}^T \left\{\Delta\hat{D}_1^1\right\}}{\left\{\Delta\hat{D}_1^{i-1}\right\}^T \left\{\Delta\hat{D}_1^i\right\}} \tag{59}$$

For the iterative step $(j \geq 2)$, the load increment parameter λ_j^i is calculated as

$$\lambda_j^i = -\frac{\left\{\Delta\hat{D}_1^{i-1}\right\}^T \left\{\Delta\bar{D}_j^i\right\}}{\left\{\Delta\hat{D}_1^{i-1}\right\}^T \left\{\Delta\hat{D}_j^i\right\}} \tag{60}$$

where $\left\{\Delta\hat{D}_1^{i-1}\right\}$ is the displacement increment generated by the reference load at the first iteration of the previous incremental step; and $\left\{\Delta\hat{D}_j^i\right\}$ and $\left\{\Delta\bar{D}_j^i\right\}$ denote the displacement increments generated by the reference load and unbalanced force vectors, respectively, at the j th iteration of the i th incremental step, as defined in Eqs. (52) and (53).

3. Numerical examples

In this section, three numerical examples are presented to verify the accuracy and efficiency of two proposed analysis methods: (1) the refined plastic hinge method and (2) the fiber method. The predictions of strength and load-displacement relationship are compared with those generated by commercial finite element packages and other existing solutions. The first example is to show how the stability functions capture the $P-\delta$ effect accurately and efficiently. The second one is to show how well the refined plastic hinge model and fiber hinge model predict the strength and behavior of frames. The last one is to demonstrate the capability of two proposed methods in predicting the strength and behavior of a large-scale twenty-story space frame. Five integration points along the length of a member and eighty fibers on the cross-section are used in the fiber model.

3.1. Elastic buckling of columns

The aim of this example is to show the accuracy and efficiency of the stability functions in capturing the elastic buckling loads of columns with different boundary conditions. Fig. 11 shows cantilever and simply supported columns. The section of columns is W8×31. The Young's modulus and Poisson ratio of the material are $E = 200{,}000$ MPa and $\nu = 0.3$, respectively. The buckling load of the columns is obtained using the load-deflection analysis. The geometric imperfection is modeled by equivalent notional lateral loads as shown in Fig. 11.

Fig. 12 shows the load-displacement curves of the columns predicted by the present element and the cubic frame element of SAP2000. Since the present element is based on the stability functions which are derived from the closed-form solution of a beam-column subjected to end forces, it can accurately predict the buckling load of columns with different boundary conditions by using only one element per member. Whereas the cubic frame element of

SAP2000, which is based on the cubic interpolation functions, overpredicts the buckling loads by 18% and 16% for the cantilever column and simply supported column, respectively, when the columns are modeled by one element per member. The load-displacement curves shown in Fig. 12 indicate that SAP2000 requires more than five cubic elements per member in modeling to match the results predicted by the present element. This is due to the fact that when the member is divided into many elements, the $P-\delta$ effect is transformed to the $P-\Delta$ effect, and hence, the results of cubic element are close to the obtained results.

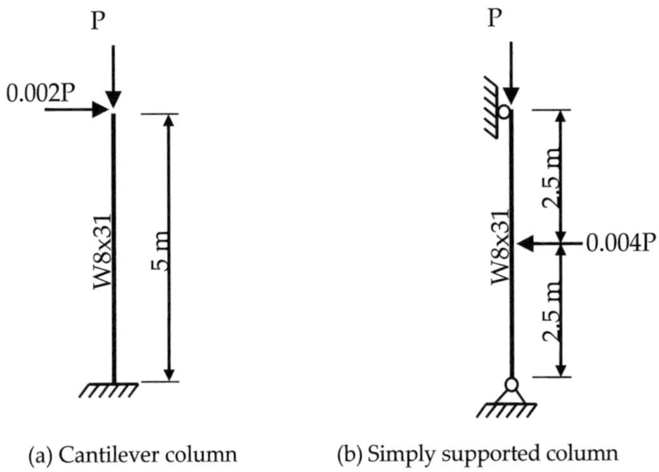

Figure 11. Steel columns

3.2. Two story space frame

A two-story space subjected to combined action of gravity load and lateral load is depicted in Fig. 13 with its geometric dimension. The Young modulus, Poisson ratio, and yield stress of material are $E = 19{,}613$ MPa, $\nu = 0.3$, and $\sigma_y = 98$ MPa, respectively. This frame was previously analyzed by De Souza (2000) using the force-based method with fiber model. De Souza (2000) used one element per member in the modeling. The B23 element of ABAQUS is also employed to model this frame. Each framed member is modeled by one present element. The aim of this example is to demonstrate capability of the present element in capturing the effects of both geometric and material nonlinearities.

The ultimate loads of the frame obtained by different methods are presented in Table 1. The load-displacement responses of the frame are also plotted in Fig. 14. It can be seen that the results of the present element are well compared with those of De Souza (2000) using the force-based method. It should be noted that only one element per member is used in present study and De Souza (2000). The B23 element of ABAQUS overestimates ultimate strength of this frame if each framed member is modeled by less than fifty B23 elements. The difference between B23 element and present element is negligible when more than fifty B32 elements are used, and the ultimate strength and load-displacement curve obtained by ABAQUS and present study are then close each other.

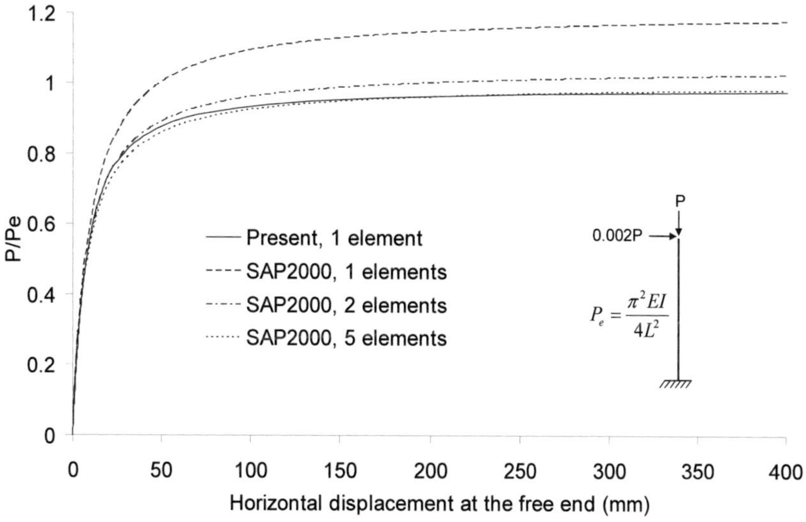

(a) Cantilever column

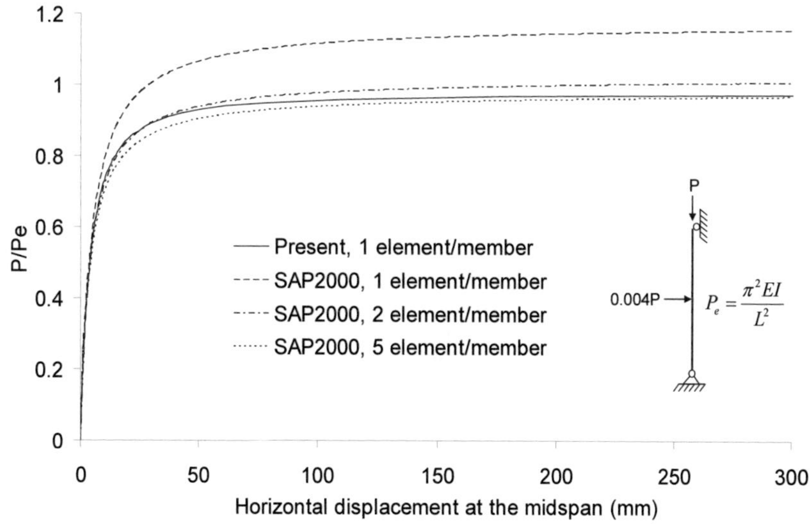

(b) Simply supported column

Figure 12. Load-displacement curves of steel columns

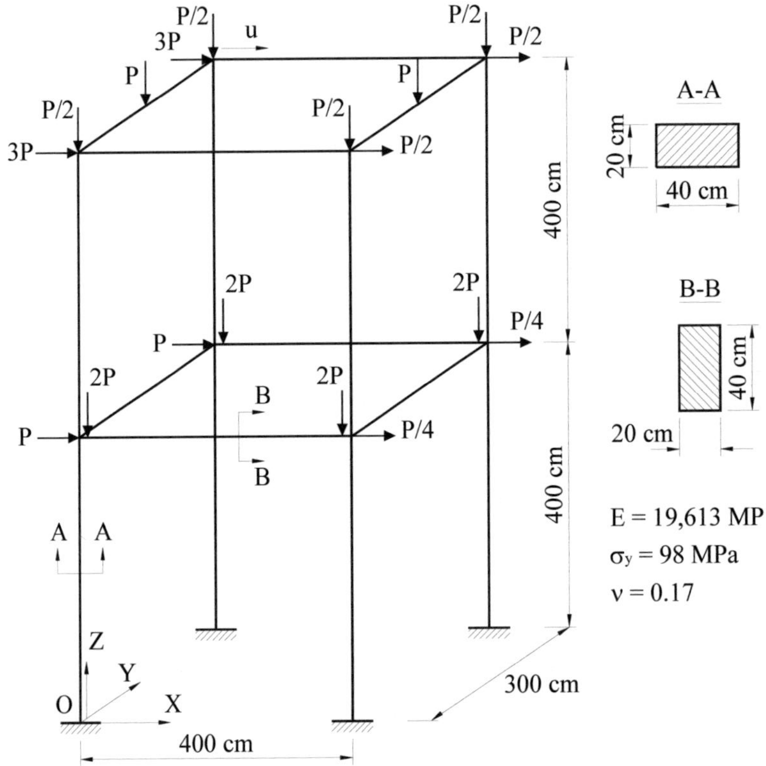

Figure 13. Two-story space frame

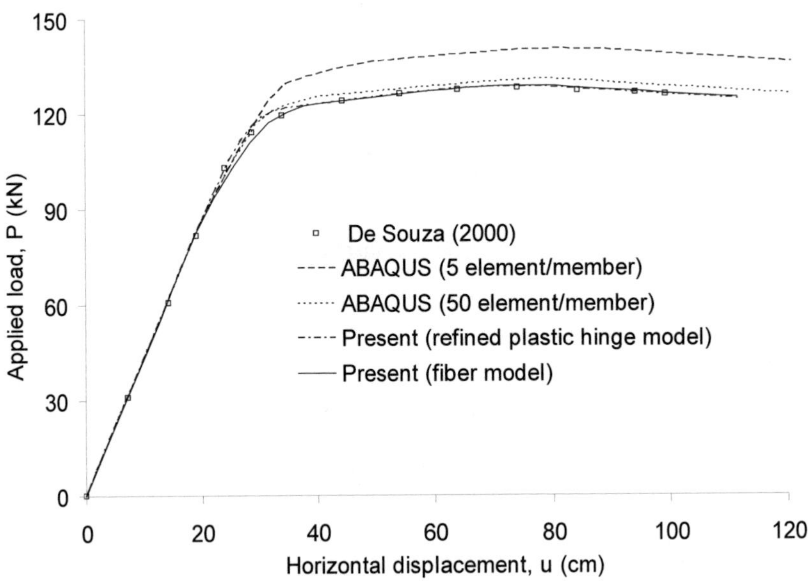

Figure 14. Load- displacement curves of two-story space frame

Method	Ultimate load (kN)	Difference (%)
De Souza (2000)	128.05	-
ABAQUS (5 element/member)	140.26	9.53
ABAQUS (20 element/member)	132.19	3.23
ABAQUS (50 element/member)	130.74	2.10
Present (refined plastic hinge model)	128.50	0.35
Present (fiber model)	128.82	0.60

Table 1. Comparison of ultimate load of two-story space frame

3.3. Twenty-story space frame

The last example is a large scale twenty-story space steel frame as shown in Fig. 15. The aim of this example is to demonstrate the capability of two proposed methods in predicting the strength and behavior of large-scale structures. A50 steel with yield stress of 344.8 Mpa, Young's modulus of 200 Gpa, and Poisson's ratio of 0.3 is used for all sections. The load applied to the structure consists of gravity loads of 4.8 kN/m^2 and wind loads of 0.96 kN/m^2 acting in the Y-direction. These loads are converted into concentrated loads applied at the beam-column joints. The obtained results are also compared with those generated by Jiang et al. (2002) using the mixed element method.

Jiang et al. (2002) used both the plastic hinge and spread-of-plasticity elements to model this structure to shorten the computational time because the use of a full spread-of-plasticity analysis is very computationally intensive. When a member modeling by one plastic hinge element detected yielding to occur between the two ends, it was divided into eight spread-of-plasticity elements to accurately capture the inelastic behavior. In this study, each framed member is modeled by only one proposed element. The load-displacement curves of node A at the roof of the frame obtained by the present elements and mixed element of Jiang et al. (2002) are shown in Fig. 16. The ultimate load factor of the frame is also given in Table 2. A very good agreement between the results is seen.

Method	Ultimate load factor	Difference (%)
Jiang et al. (2002)	1.000	-
Present (refined plastic hinge model)	1.021	2.10
Present (fiber model)	1.0002	0.02

Table 2. TAnalysis result of twenty-story space frame

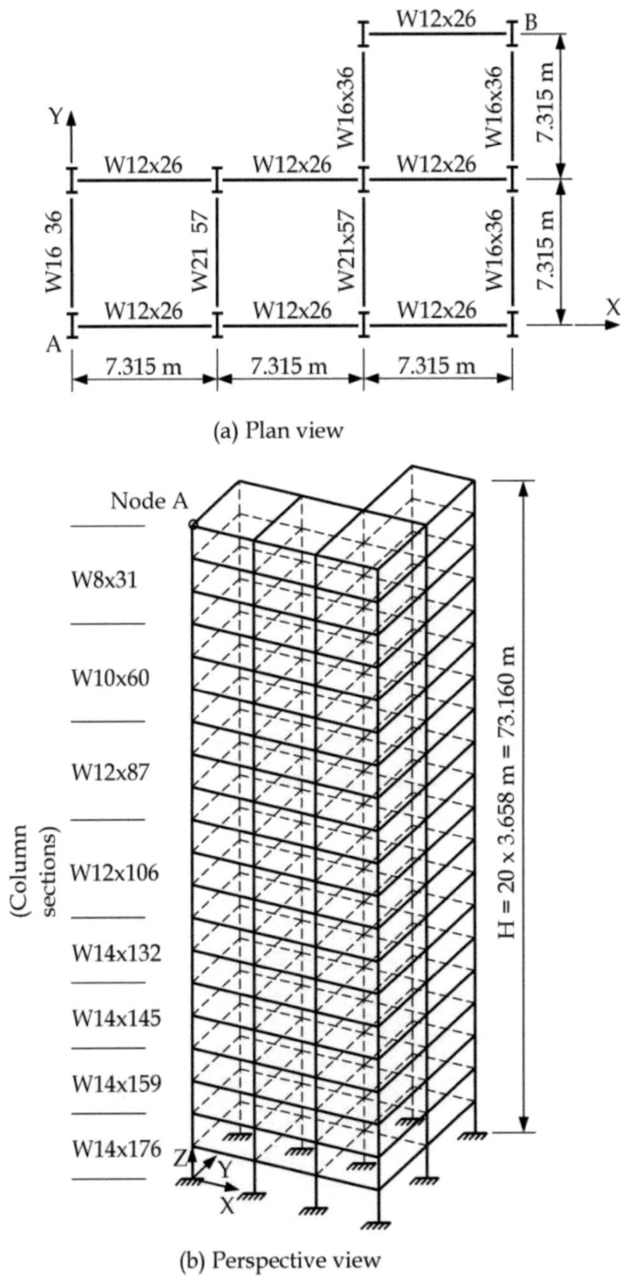

Figure 15. Twenty-story space frame

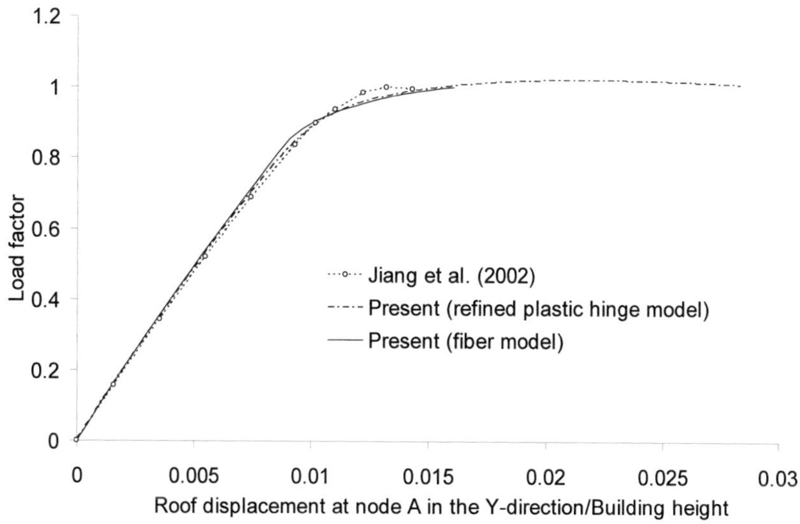

Figure 16. Load-displacement curves of twenty-story space frame

4. Conclusion

This chapter has presented two advanced analysis methods for space steel frames. In these methods, the geometric nonlinearities are captured using the stability functions, while the material nonlinearities are considered using the refined plastic hinge model and fiber model. The benefit of using the stability functions is that they require only one element per member, and hence, minimize the modeling and solution time. The advantage of refined plastic hinge model is its simplicity and efficiency. However, it is limited to steel material. Although the fiber model is a little bit time consuming compared to the refined plastic hinge model, it can be used for both steel and concrete or concrete-filled steel tubular structures as shown in the works of Thai & Kim (2011a).

Author details

Huu-Tai Thai
Hanyang University, South Korea

5. References

[1] AISC. (2005). *Load and resistance factor design specification for structural steel buildings,* American Institute of Steel Construction, Chicago, Illinois.

[2] Batoz, J. L. & Dhatt, G. (1979). Incremental displacement algorithms for nonlinear problems. *International Journal for Numerical Methods in Engineering,* Vol. 14, No. 8, pp. 1262-1297.

[3] De Souza, R. (2000). Force-based finite element for large displacement inelastic analysis of frames. PhD Dissertation, *Department of Civil and Environmental Engineering, University of California at Berkeley.*

[4] Jiang, X. M.; Chen, H. & Liew, J. Y. R. (2002). Spread-of-plasticity analysis of three-dimensional steel frames. *Journal of Constructional Steel Research*, Vol. 58, No. 2, pp. 193-212.

[5] McGuire, W.; Ziemian, R. D. & Gallagher, R. H. (2000). *Matrix structural analysis*, John Wiley & Sons, New York.

[6] Spacone, E.; Filippou, F. & Taucer, F. (1996). Fibre beam-column model for non-linear analysis of R/C frames: part I. Formulation. *Earthquake Engineering and Structural Dynamics*, Vol. 25, No. 7, pp. 711-725.

[7] Thai, H. T. & Kim, S. E. (2008). Second-order inelastic dynamic analysis of three-dimensional cable-stayed bridges. *International Journal of Steel Structures*, Vol. 8, No. 3, pp. 205-214.

[8] Thai, H. T. & Kim, S. E. (2009). Practical advanced analysis software for nonlinear inelastic analysis of space steel structures. *Advances in Engineering Software*, Vol. 40, No. 9, pp. 786-797.

[9] Thai, H. T. & Kim, S. E. (2011a). Nonlinear inelastic analysis of concrete-filled steel tubular frames. *Journal of Constructional Steel Research*, Vol. 67, No. 12, pp. 1797-1805.

[10] Thai, H. T. & Kim, S. E. (2011b). Nonlinear inelastic analysis of space frames. *Journal of Constructional Steel Research*, Vol. 67, No. 4, pp. 585-592.

[11] Thai, H. T. & Kim, S. E. (2011c). Practical advanced analysis software for nonlinear inelastic dynamic analysis of steel structures. *Journal of Constructional Steel Research*, Vol. 67, No. 3, pp. 453-461.

[12] Thai, H. T. & Kim, S. E. (2011d). Second-order inelastic dynamic analysis of steel frames using fiber hinge method. *Journal of Constructional Steel Research*, Vol. 67, No. 10, pp. 1485-1494.

[13] Thai, H. T. & Kim, S. E. (2012). Second-order inelastic analysis of cable-stayed bridges. *Finite Elements in Analysis and Design*, Vol. 53, No. 6, pp. 48-55.

[14] Yang, Y. B. & Shieh, M. S. (1990). Solution method for nonlinear problems with multiple critical points. *AIAA Journal*, Vol. 28, No. 12, pp. 2110-2116.

Chapter 5

Analytical, Numerical and Experimental Studies on Stability of Three-Segment Compression Members with Pinned Ends

Seval Pinarbasi Cuhadaroglu, Erkan Akpinar,
Fuad Okay, Hilal Meydanli Atalay and Sevket Ozden

Additional information is available at the end of the chapter

http://dx.doi.org/10.5772/45807

1. Introduction

In earthquake resistant structural steel design, there are two commonly used structural systems. "Moment resisting frames" consist of beams connected to columns with moment resisting (i.e., rigid) connections. Rigid connection of a steel beam to a steel column requires rigorous connection details. On the other hand, in "braced frames", the simple (i.e., pinned) connections of beams to columns are allowed since most of the earthquake forces are carried by steel braces connected to joints or frame elements with pinned connections. The load carrying capacity of a braced frame almost entirely based on axial load carrying capacities of the braces. If a brace is under tension in one half-cycle of an earthquake excitation, it will be subjected to compression in the other half cycle. Provided that the connection details are designed properly, the tensile capacity of a brace is usually much higher than its compressive capacity. In fact, the fundamental limit state that governs the behavior of such steel braces under seismic forces is their global buckling behavior under compression.

After detailed evaluation, if a steel braced structure is decided to have insufficient lateral strength/stiffness, it has to be strengthened/stiffened, which can be done by increasing the load carrying capacities of the braces. The key parameter that controls the buckling capacity of a brace is its "slenderness" (Salmon et al., 2009). As the slenderness of a brace decreases, its buckling capacity increases considerably. In order to decrease the slenderness of a brace, either its length has to be decreased, which is usually not possible or practical due to architectural reasons, or its flexural stiffness has to be increased. Flexural stiffness of a brace can be increased by welding steel plates or by wrapping fiber reinforced polymers around the steel section. Analytical studies (e.g., Timoshenko & Gere, 1961) have shown that it usually leads to more economic designs if only the partial length, instead of the entire length, of the brace is stiffened. This also eliminates possible complications in connection details that have to be considered at the ends of the member.

Nonuniform structural elements are not only used in seismic strengthening and rehabilitation of existing structures. In an attempt to design economic and aesthetic structures, many engineers and architects nowadays prefer to use nonuniform structural elements in their structural designs. However, stability analysis of such nonuniform members is usually much more complex than that of uniform members (e.g., see Li, 2001). In fact, most of the design formulae/charts given in design specifications are developed for uniform members. Thus, there is a need for a practical tool to analyze buckling behavior of nonuniform members.

This study investigates elastic buckling behavior of three-segment symmetric stepped compression members with pinned ends (Fig. 1) using three different approaches: (i) analytical, (ii) numerical and (iii) experimental approaches. As already mentioned, such a member can easily be used to strengthen/rehabilitate an existing steel braced frame or can directly be used in a new construction. Surely, the use of stepped elements is not only limited to the structural engineering applications; they can be used in many other engineering applications, such as in mechanical and aeronautical engineering.

In analytical studies, first the governing equations of the studied stability problem are derived. Then, exact solution to the problem is obtained. Since exact solution requires finding the smallest root of a rather complex characteristic equation which highly depends on initial guess, the governing equation is also solved using a recently developed analytical technique by He (1999), which is called Variational Iteration Method (VIM). Many researchers (e.g., Abulwafa et al., 2007; Batiha et al., 2007; Coskun & Atay, 2007, 2008; Ganji & Sadighi, 2007; Miansari et al., 2008; Ozturk, 2009 and Sweilan & Khader, 2007) have shown that complex engineering problems can easily and successfully be solved using VIM. Recently, VIM has also been applied to stability analysis of compression and flexural members. Coskun and Atay (2009), Atay and Coskun (2009), Okay et al. (2010) and Pinarbasi (2011) have shown that it is much easier to solve the resulting characteristic equation derived using VIM. In this paper, by comparing the approximate VIM results with the exact results, the effectiveness of using VIM in determining buckling loads of multi-segment compression members is investigated.

The problem is also handled, for some special cases, using widely known structural analysis program SAP2000 (CSI, 2008). After determining the buckling load of a uniform member with a hollow rectangular cross section, the stiffness of the member is increased along its length partially in different length ratios and the effect of such stiffening on buckling load of the member is investigated. By comparing numerical results with analytical results, the effectiveness of using such an analysis program in stability analysis of multi-segment elements is also investigated.

Finally, buckling loads of uniform and three-segment stepped steel compression members with hollow rectangular cross section are determined experimentally. In the experiments, the "stiffened" columns are prepared by welding additional steel plates over two sides of the member in such a way that the addition of the plates predominantly increases the smaller flexural rigidity of the cross section, which governs the buckling behavior of the member. By changing the length of the stiffening plates, i.e., by changing the stiffened length ratio, the degree of overall stiffening is investigated in the experimental study. The experimental study also shows in what extent the *ideal* conditions assumed in analytical and numerical studies can be realized in a laboratory research.

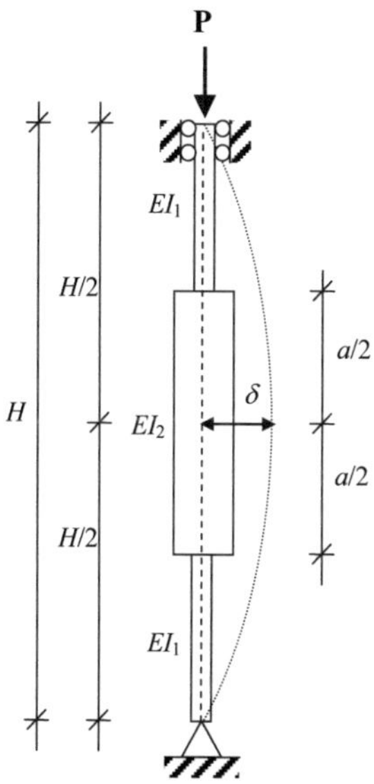

Figure 1. Three-segment symmetric stepped compression member with pinned ends

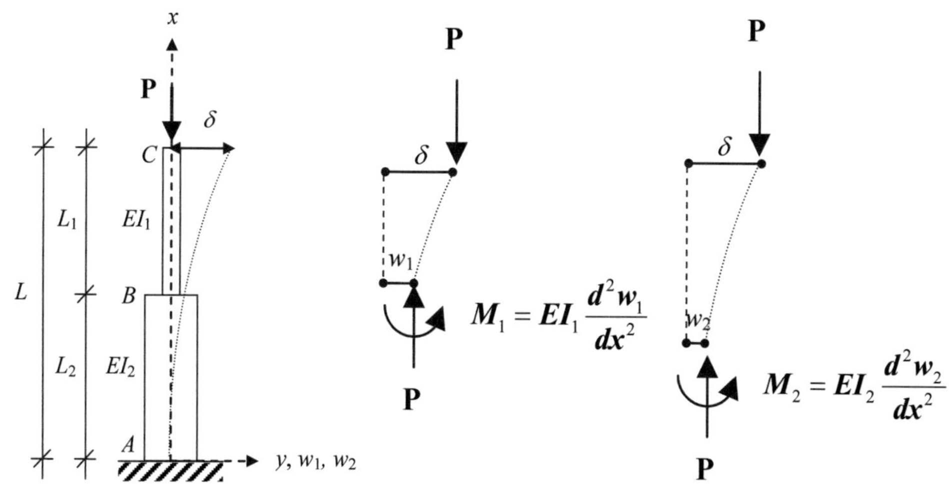

a. undeformed and deformed shapes

b. free body diagram for Segment I

c. free body diagram for Segment II

Figure 2. "Equivalent" two-segment stepped compression member with one end fixed (clamped), the other hinged

2. Analytical studies on elastic buckling of a three-segment stepped compression member with pinned ends

2.1. Derivation of governing (buckling) equations

Consider a three-segment symmetric stepped compression member subjected to a compressive load P applied at its top end, as shown in Fig. 1. Assume that both ends of the member are pinned; i.e., free to rotate. Also assume that the top and bottom segments of the member have identical flexural stiffness, EI_1, while that of the middle segment may be different, say EI_2. As long as the stiffness variation along the height of the member is symmetric about the mid-height, the buckled shape of the member is also symmetric about the same point as shown in Fig. 1. When such a symmetry exists, the buckling load of the three-segment member can be obtained by analyzing the simpler two-segment member shown in Fig. 2a. This "equivalent" two-segment member has a fixed (clamped) boundary condition at its bottom end whereas its top end is free. From comparison of Fig. 1 and Fig. 2a, one can also see that the length of the equivalent two-segment member equals to the half-length of the original three-segment member, i.e., $L=H/2$. Similarly, $L_2=a/2$. Since the analysis of a two-segment column is much simpler than that of a three-segment column, the analytical study presented in this section is based on the equivalent two-segment member.

The undeformed and deformed shapes of the equivalent two-segment member under uniform compression are illustrated in Fig. 2a. The origin of x-y coordinate system is located at the bottom end of the column. Since the stiffnesses of two segments of the column can be different in general, each segment of the column has to be analyzed separately. Equilibrium equation at an arbitrary section in Segment I can be written from the free body diagram shown in Fig. 2b:

$$EI_1 \frac{d^2 w_1}{dx^2} - P(\delta - w_1) = 0 \tag{1}$$

which can be expressed as

$$\frac{d^2 w_1}{dx^2} + k_1^2 w_1 = k_1^2 \delta \text{ where } k_1^2 = \frac{P}{EI_1} \tag{2}$$

In Eq. (1) and Eq. (2), w_1 is lateral displacement of Segment I at any point, δ is the lateral displacement of the top end of the member, i.e., $\delta = w_1\ (x = L)$. Eq. (2) is valid for $L_2 \le x \le L$. Similarly, from Fig. 2c, the equilibrium equation at an arbitrary section in Segment II can be written as

$$\frac{d^2 w_2}{dx^2} + k_2^2 w_2 = k_2^2 \delta \text{ where } k_2^2 = \frac{P}{EI_2} \tag{3}$$

where w_2 is the displacement of Segment II in y direction. Eq. (3) is valid for $0 \le x \le L_2$. For easier computations, the buckling equations in Eq. (2) and Eq. (3) can be written in nondimensional form as follows:

$$(\bar{w}_1)'' + \beta_1^2(\bar{w}_1) = \beta_1^2(\bar{\delta}) \text{and} \quad (\bar{w}_2)'' + \beta_2^2(\bar{w}_2) = \beta_2^2(\bar{\delta}) \tag{4}$$

with

$$\beta_1 = k_1 L \text{and} \quad \beta_2 = k_2 L \tag{5}$$

where $\bar{x} = x / L$, $\bar{w}_1 = w_1 / L$, $\bar{w}_2 = w_2 / L$, $\bar{\delta} = \delta / L$ and prime denotes differentiation with respect to $\bar{x}$. Since both of the differential equations in Eq. (4) are in second order, the solutions will contain four integration constants. Considering that δ is also unknown, the solution of these buckling equations requires five conditions to determine the resulting five unknowns. Two of these conditions come from the continuity conditions where the flexural stiffness of the column changes and the remaining three conditions are obtained from the boundary conditions at the ends of the column. At $x=L_2$, the lateral displacement and slope functions have to be continuous, which requires

$$[\bar{w}_1]_{\bar{x}=s} = [\bar{w}_2]_{\bar{x}=s} \text{and} \quad \left[(\bar{w}_1)'\right]_{\bar{x}=s} = \left[(\bar{w}_2)'\right]_{\bar{x}=s} \tag{6}$$

where $s = L_2 / L$. As far as the boundary conditions are concerned, for a clamped-free column, the end conditions can be written in nondimensional form as:

$$\left[(\bar{w}_2)\right]_{\bar{x}=0} = 0, \quad \left[(\bar{w}_2)'\right]_{\bar{x}=0} = 0 \text{and} \quad \left[(\bar{w}_1)\right]_{\bar{x}=1} = \bar{\delta} \tag{7}$$

Thus, Eq. (4) with Eq. (6) and Eq. (7) constitutes the governing equations for the studied stability problem.

2.2. Exact solution to buckling equations

Since the differential equations given in Eq. (4) are relatively simple, it is not too difficult to obtain their exact solutions, which can be written in the following form:

$$\bar{w}_1 = C_1 \sin(\beta_1 \bar{x}) + C_2 \cos(\beta_1 \bar{x}) + \bar{\delta} \text{and} \quad \text{and} \quad \bar{w}_2 = C_3 \sin(\beta_2 \bar{x}) + C_4 \cos(\beta_2 \bar{x}) + \bar{\delta} \tag{8}$$

where C_i (i=1-4) are integration constants to be determined from continuity and end conditions. From the first and second conditions given in Eq. (7), one can find that

$$C_3 = 0 \text{and} \quad \text{and} \quad C_4 = -\bar{\delta} \tag{9}$$

Then, using Eq. (6), the other integration constants are obtained as:

$$C_1 = \bar{\delta}\left[\frac{\beta_2}{\beta_1}\sin(\beta_2 s)\cos(\beta_1 s) - \cos(\beta_2 s)\sin(\beta_1 s)\right] \tag{10a}$$

$$C_2 = -\bar{\delta}\left[\frac{\beta_2}{\beta_1}\sin(\beta_2 s)\sin(\beta_1 s) + \cos(\beta_2 s)\cos(\beta_1 s)\right] \tag{10b}$$

Finally, the last condition given in Eq. (7) results in

$$\left\{\tan\left[\beta_2 s\right]\tan\left[\beta_1\left(1-s\right)\right]-\frac{\beta_1}{\beta_2}\right\}\bar{\delta}=0 \tag{11}$$

For a nontrivial solution, the coefficient term must be equal to zero, yielding the following characteristic equation for the studied buckling problem:

$$\tan\left[\beta_2 s\right]\tan\left[\beta_1\left(1-s\right)\right]=\frac{\beta_1}{\beta_2} \tag{12}$$

Since $\beta_1 / \beta_2 = \sqrt{EI_2 / EI_1}$, if the stiffness ratio n is defined as $n = EI_2 / EI_1$, Eq. (12) can be written in terms of β_1 (square root of nondimensional buckling load of the equivalent two-segment element in terms of EI_1), n (stiffness ratio) and s (stiffened length ratio) as follows:

$$\tan\left[\beta_1\left(1-s\right)\right]\tan\left[\beta_1\frac{s}{\sqrt{n}}\right]=\sqrt{n} \tag{13}$$

One can show that the buckling load of the three-segment stepped compression member with length H shown in Fig. 1 can be written in terms of that of the equivalent two-segment member with length L=H/2 shown in Fig. 2a as

$$P_{cr}=\lambda\frac{EI_1}{H^2}\text{ where }\lambda=4\beta_1^2 \tag{14}$$

In other words, λ is the nondimensional buckling load of the three-segment compression member *in terms of* EI_1.

2.3. VIM solution to buckling equations

According to the variational iteration method (VIM), a general nonlinear differential equation can be written in the following form:

$$Lw\left(x\right)+Nw\left(x\right)=g\left(x\right) \tag{15}$$

where L is a linear operator and N is a nonlinear operator, $g(x)$ is the nonhomogeneous term. Based on VIM, the "correction functional" can be constructed as

$$w_{n+1}\left(x\right)=w_n\left(x\right)+\int_0^x\lambda\left(\xi\right)\left\{Lw_n\left(\xi\right)+N\tilde{w}_n\left(\xi\right)\right\}d\xi \tag{16}$$

where $\lambda\left(\xi\right)$ is a general Lagrange multiplier that can be identified optimally via variational theory, w_n is the n-th approximate solution and $\tilde{w}_n$ denotes a restricted variation, i.e., $\delta\tilde{w}_n=0$ (He, 1999). As summarized in He et al. (2010), for a second order differential equation such as the buckling equations given in Eq. (4), $\lambda\left(\xi\right)$ simply equals to

$$\lambda(\xi) = (\xi - x) \tag{17}$$

The original variational iteration algorithm proposed by He (1999) has the following iteration formula:

$$w_{n+1}(x) = w_n(x) + \int_0^x \lambda(\xi)\{Lw_n(\xi) + Nw_n(\xi)\}d\xi \tag{18}$$

In a recent paper, He et al. (2010) proposed two additional variational iteration algorithms for solving various types of differential equations. These algorithms can be expressed as follows:

$$w_{n+1}(x) = w_0(x) + \int_0^x \lambda(\xi)\{Nw_n(\xi)\}d\xi. \tag{19}$$

and

$$w_{n+2}(x) = w_{n+1}(x) + \int_0^x \lambda(\xi)\{Nw_{n+1}(\xi) - Nw_n(\xi)\}d\xi \tag{20}$$

Thus, the three VIM iteration algorithms for the buckling equations given in Eq. (4) can be written as follows:

$$\overline{w}_{i,n+1}(x) = \overline{w}_{i,n}(x) + \int_0^x (\xi - x)\left\{\overline{w}''_{i,n}(\xi) + \beta_i^2 \overline{w}_{i,n} - \beta_i^2 \overline{\delta}\right\}d\xi, \tag{21a}$$

$$\overline{w}_{i,n+1}(x) = \overline{w}_{i,0}(x) + \int_0^x (\xi - x)\left\{\beta_i^2 \overline{w}_{i,n} - \beta_i^2 \overline{\delta}\right\}d\xi, \tag{21b}$$

$$\overline{w}_{i,n+2}(x) = \overline{w}_{i,n+1}(x) + \int_0^x (\xi - x)\left\{\left(\overline{w}''_{i,n+1}(\xi) - \overline{w}''_{i,n}(\xi)\right) + \beta_i^2\left(\overline{w}_{i,n+1} - \overline{w}_{i,n}\right)\right\}d\xi, \tag{21c}$$

where i is the segment number and can take the values of one or two. It has already been shown in Pinarbasi (2011) that all VIM algorithms yield exactly the same results for a similar stability problem. For this reason, considering its simplicity, the second iteration algorithm is decided to be used in this study.

Recalling that $\beta_1 / \beta_2 = \sqrt{n}$ and $\lambda = 4\beta_1^2$, the iteration formulas for the buckling equations of the studied problem can be written in terms of λ and n as follows:

$$\overline{w}_{1,j+1}(x) = \overline{w}_{1,0}(x) + \int_0^x (\xi - x)\left\{\frac{\lambda}{4}\left(\overline{w}_{1,j} - \overline{\delta}\right)\right\}d\xi, \tag{22a}$$

$$\overline{w}_{2,j+1}(x)=\overline{w}_{2,0}(x)+\int_0^x(\xi-x)\left\{\frac{\lambda}{4n}\left(\overline{w}_{2,j}-\overline{\delta}\right)\right\}d\xi \tag{22b}$$

As an initial approximation for displacement function of each segment, a linear function with unknown coefficients is used:

$$\overline{w}_{1,0}=C_1\overline{x}+C_2 \text{ and } \overline{w}_{2,0}=C_3\overline{x}+C_4 \tag{23}$$

where C_i (i=1-4) are to be determined from continuity and end conditions. After conducting seventeen iterations, $\overline{w}_{1,17}$ and $\overline{w}_{2,17}$ are obtained. Substituting these approximate solutions to the continuity equations in Eq. (6) and to the end conditions in Eq. (7), five equations are obtained. Four of them are used to determine the unknown coefficients in terms of $\overline{\delta}$, while the remaining one is used to construct the characteristic equation for the studied problem:

$$\left[F(\lambda)\right]\overline{\delta}=0 \tag{24}$$

where $F(\lambda)$ is the coefficient term of $\overline{\delta}$. For a nontrivial solution $F(\lambda)$ must be equal to zero. The smallest possible real root of the characteristic equation gives the nondimensional buckling load ($\lambda=PH^2/EI_1$) of the three-segment compression member in the first buckling mode.

2.4. Comparison of VIM results with exact results

For various values of stiffness ratio (n=EI_2/EI_1) and stiffened length ratio (s=a/H), nondimensional buckling loads of a three-segment compression member with pinned ends are determined both by using Eq. (13) and VIM. VIM results are compared with the exact results in Table 1.

n	s							
	0.2		0.4		0.6		0.8	
	Exact	VIM	Exact	VIM	Exact	VIM	Exact	VIM
100	15.344	15.344	27.052	27.052	59.843	59.843	225.706	225.706
10	14.675	14.675	24.006	24.006	44.978	44.978	85.880	85.880
5	13.978	13.978	21.109	21.109	33.471	33.471	46.651	46.651
2.5	12.721	12.721	16.694	16.693	21.275	21.275	24.186	24.186
1.67	11.632	11.632	13.642	13.642	15.406	15.406	16.306	16.306
1.25	10.689	10.689	11.471	11.471	12.039	12.039	12.297	12.297

Table 1. Comparison of VIM predictions for nondimensional buckling load (λ) of a three-segment compression member with exact results for various values of stiffness ratio (n=EI_2/EI_1) and stiffened length ratio (s=a/H)

As it can be seen from Table 1, VIM results perfectly match with exact results, verifying the efficiency of VIM in this particular stability problem. It is worth noting that it is somewhat difficult to solve the characteristic equation given in Eq. (13) since it is highly sensitive to the initial guess. While solving this equation, one should be aware of that an improper initial guess can result in a buckling load in higher modes. On the other hand, the characteristic equations derived using VIM are composed of polynomials, all roots of which can be obtained more easily. This is one of the strength of VIM even when an exact solution is available for the problem, as in our case.

2.5. VIM results for various stiffness and stiffened length ratios

Table 2 tabulates VIM predictions for nondimensional buckling load of a three-segment stepped compression member for various values of stiffness (n) and stiffened length (s) ratios. The results listed in this table can directly be used by design engineers who design/strengthen three-segment symmetric stepped compression members with pinned ends.

n	**s**						
	0.1	**0.2**	**0.25**	**0.3333**	**0.5**	**0.75**	**0.9999**
1	9.8696	9.8696	9.8696	9.8696	9.8696	9.8696	9.8696
1.5	10.5592	11.3029	11.6881	12.3342	13.5322	14.6186	14.8044
2	10.9332	12.1571	12.8290	14.0255	16.5379	19.2404	19.7392
2.5	11.1676	12.7211	13.6051	15.2433	19.0149	23.7328	24.6740
3	11.3282	13.1202	14.1651	16.1557	21.0707	28.0942	29.6088
4	11.5338	13.6465	14.9165	17.4239	24.2442	36.4193	39.4784
5	11.6599	13.9775	15.3962	18.2587	26.5469	44.2105	49.3480
7.5	11.8311	14.4372	16.0711	19.4641	30.1728	61.3848	74.0220
10	11.9181	14.6750	16.4240	20.1076	32.2453	75.4700	98.6960
20	12.0504	15.0419	16.9731	21.1249	35.6828	109.4880	197.3920
50	12.1307	15.2680	17.3139	21.7652	37.9220	138.1940	493.4800
100	12.1577	15.3444	17.4295	21.9836	38.6944	148.2010	986.9600

Table 2. VIM predictions for nondimensional buckling load (λ) of a three-segment column for various values of stiffness ratio ($n=EI_2/EI_1$) and stiffened length ratio ($s=a/H$)

At this stage, it can be valuable to investigate the amount of increase in buckling load due to partial stiffening of a compression member. Fig. 3 shows variation of increase in critical buckling load, with respect to the uniform case, with stiffened length ratio for different values of stiffness ratio. From Fig. 3, it can be inferred that there is no need to stiffen entire length of the member to gain appreciable amount of increase in buckling load especially if n is not too large. For n=2, increase in buckling load when only half length of the member is stiffened is more than 80 % of the increase that can be gained when the entire length of the member is stiffened. Fig. 3 also shows that if n increases, to get such an enhancement in buckling load, s has to be increased. For example, when n=10, the stiffened length of the member has to be more than 75% of its entire length if similar enhancement in member behavior is required. In fact, this can be seen more easily from Fig. 4 where the increase in buckling load is plotted in terms of stiffness ratio for various stiffened length ratios. Fig. 4 shows that if the stiffened length ratio is small, there is no need to increase the stiffness ratio too much. As an example, if only one-fifth of the entire length of the member is to be stiffened, increase in buckling load when n=2 is more than 80% of that when n=10. On the other hand, if 75 % of the entire length is allowed to be stiffened, increase in buckling load when n=2 is approximately 25% of that when n=10.

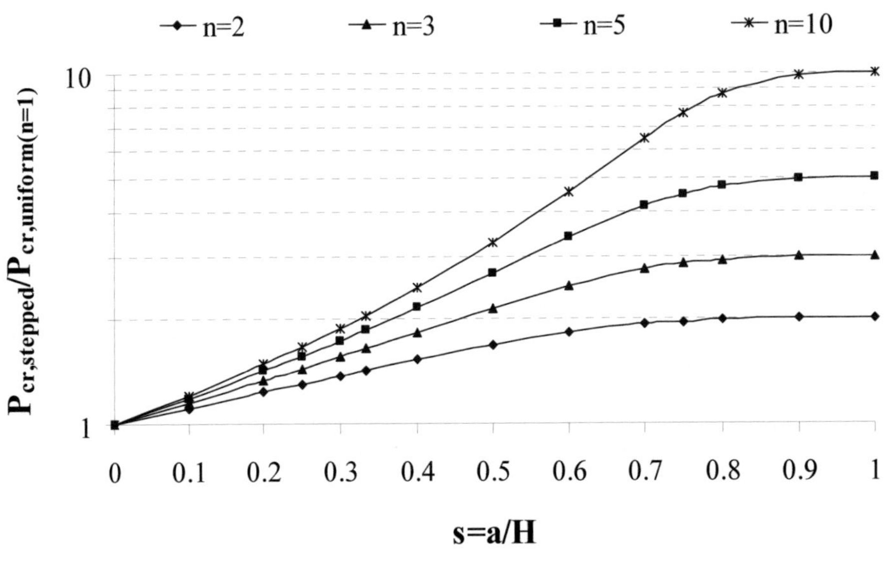

Figure 3. Variation of increase in buckling load with stiffened length ratio (s) for various values of stiffness ratio (n)

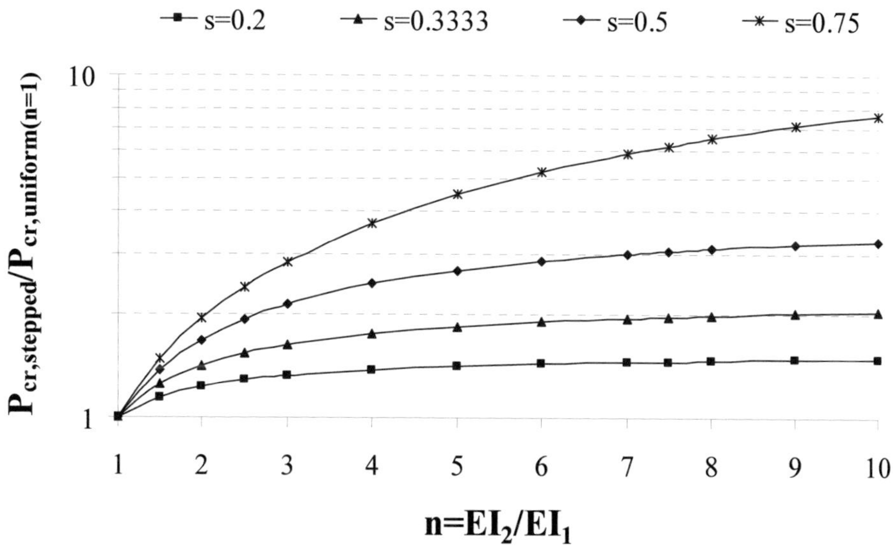

Figure 4. Variation of increase in buckling load with stiffness ratio (n) for various values of stiffened length ratio (s)

3. Numerical studies on elastic buckling of a three-segment stepped compression member with pinned ends

In order to obtain directly comparable results with the experimental results that will be discussed in the following section, in the numerical analysis, the reference "unstiffened" member is selected to have a hollow rectangular cross section, namely RCF 120x40x4, the geometric properties of which is given in Fig. 5a. The length of the steel (with modulus of elasticity of E=200 GPa) columns is chosen to be 2 m., which is the largest height of a compression member that can be tested in the laboratory due to the height limitations of the test setup. Elastic stability (buckling) analysis is performed using a well-known commercial structural analysis program SAP2000 (CSI, 2008).

Fig. 5b shows numerical solutions for the buckled shape and buckling load, $P_{cr,num,n=1}$ = 156.55 kN, of the uniform column. Exact value of the buckling load P_{cr} for this column can be computed from the well-known formula of Euler; $P_{cr} = \pi^2 EI / L^2$, which gives $P_{cr,exact,n=1}$ = 157.42 kN. The error between the numerical and exact analytical result is only 0.5 %, which encourages the use of this technique in determining the buckling load of "stiffened" members.

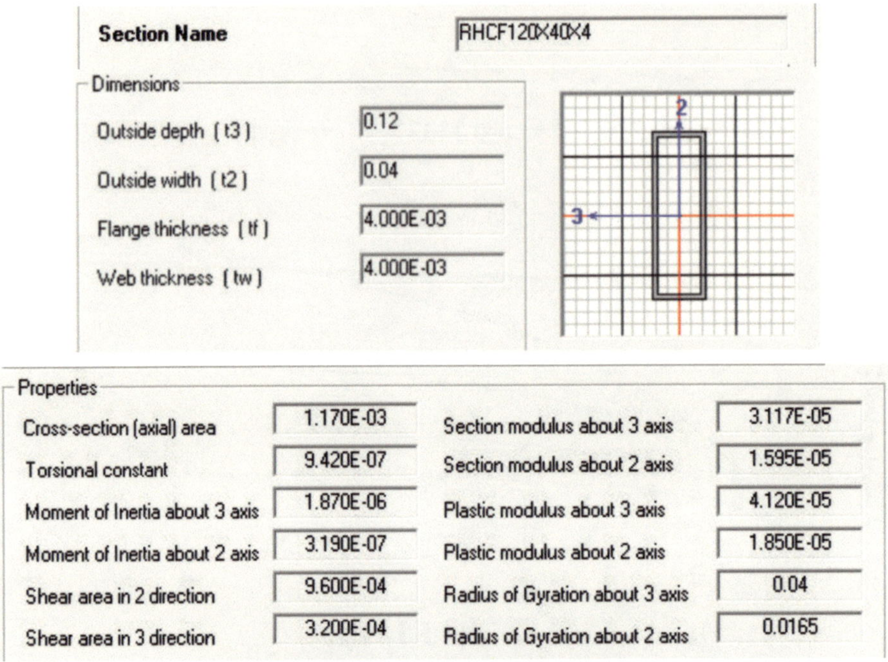

a. cross sectional properties (in meters)

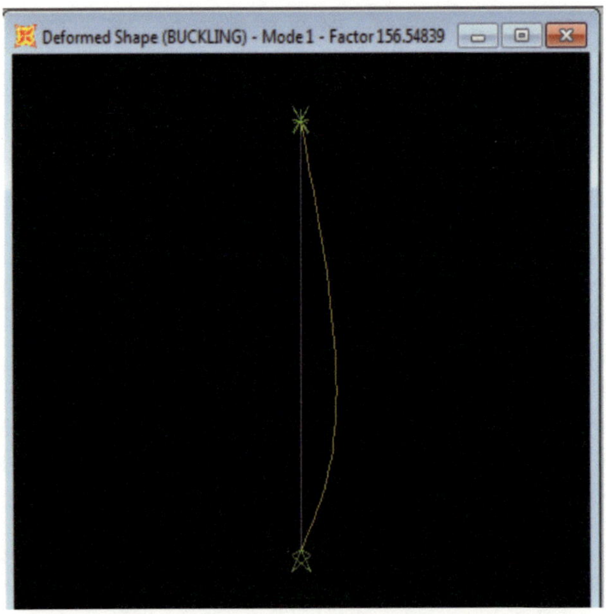

b. buckling load (in kN)

Figure 5. Geometric properties and buckling load of the uniform column (n=1) analyzed in numerical study

In the experimental study, in addition to the unstiffened members, three different types of stiffened columns are tested. In these specimens, the stiffness ratio is kept constant ($n \cong 2$) while the stiffened length ratio is varied. The stiffnesses of the three-segment members are increased by welding rectangular steel plates, with 100 mm width and 3 mm thickness as shown in Fig. 6a, to the wider faces of the hollow cross section. The length of the stiffening plates is 0.4 m in members with s=0.2, approximately 0.67 m in members with s=0.3333 and 1.0 m in members with s=0.5. This stiffening method increases the cross sectional area of the section about 1.56 times and major and minor axis flexural rigidities of the cross section, respectively, about 1.36 and 1.96 times. In the numerical analysis, the geometrical properties of the cross section for the stiffened region of the column have to be increased in these ratios. In SAP2000 (CSI, 2008), this step can easily be performed by using "property/stiffness modification factors" command (Fig. 6a). It is to be noted that axis-2 is still the minor axis of the member, so the buckling is expected to be observed about this axis, as in the uniform column case. Fig. 6b shows the buckled shape and buckling load ($P_{cr,num,n=1.96,s=0.2}$ = 192.30 kN) of the stiffened members when one-fifth of the entire length of the member is stiffened as illustrated in Fig. 6a; i.e., when n=1.96 and s=0.2. Similar analyses on members with s=0.3333 and s=0.5 yield buckling loads of $P_{cr,num,n=1.96,s=0.3333}$ = 220.42 kN and $P_{cr,num,n=1.96,s=0.5}$ = 258.93 kN, respectively. If these values of buckling loads for stiffened elements are normalized with respect to the buckling load for the uniform member ($P_{cr,num,n=1}$ = 156.55 kN), the amount of increase achieved in buckling load in each stiffening scheme is computed approximately as 1.23 when s=0.2, 1.41 when s=0.3333 and 1.65 when s=0.5. To compare numerical results with analytical results, buckling loads for three-segment symmetric stepped columns with n=1.96 are determined using VIM for various values of s and increase in buckling load with varying s is plotted in Fig. 7. It can be seen that the approximate results obtained through numerical analysis exactly match with VIM solutions. The effectiveness of the numerical analysis in solving this special buckling problem is examined further for different values of n and s. The results are presented in Table 4, which indicates very good agreement between the analytical and numerical results.

	s=0.25			s=0.5			s=0.75		
n	Exact	VIM	SAP2000	Exact	VIM	SAP2000	Exact	VIM	SAP2000
1.5	1.18	1.18	1.18	1.37	1.37	1.38	1.48	1.48	1.48
2	1.30	1.30	1.30	1.68	1.68	1.68	1.95	1.95	1.93
2.5	1.38	1.38	1.38	1.93	1.93	1.92	2.40	2.40	2.38
3	1.44	1.44	1.44	2.13	2.13	2.13	2.85	2.85	2.80
5	1.56	1.56	1.56	2.69	2.69	2.67	4.48	4.48	4.35
7.5	1.63	1.63	1.63	3.06	3.06	3.03	6.22	6.22	5.94
10	1.66	1.66	1.67	3.27	3.27	3.24	7.65	7.65	7.20

Table 3. Comparison of numerical results with analytical (exact and approximate (VIM)) results for increase in buckling load for a three-segment compression member with pinned ends for various values of stiffness ratio ($n=EI_2/EI_1$) and stiffened length ratio ($s=a/H$)

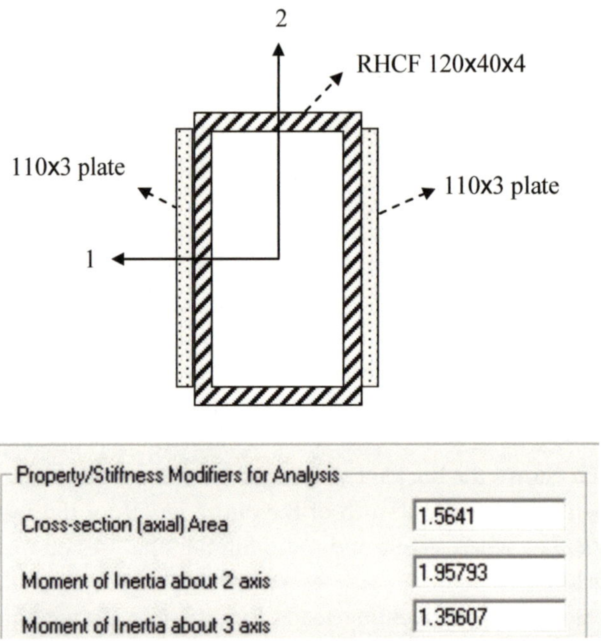

a. area/stiffness modifiers for the stiffened region of the column

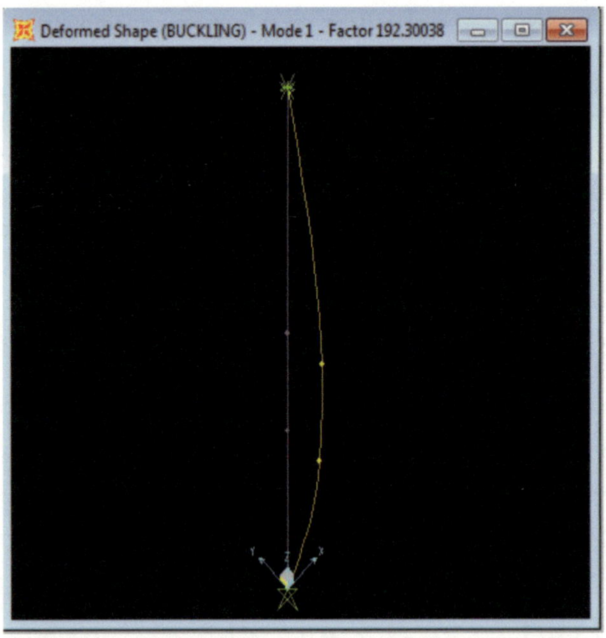

b. buckling load (in kN)

Figure 6. Geometric properties and buckling load a three-segment stepped column with stiffened length ratio s=0.2 and stiffness ratio n=1.96

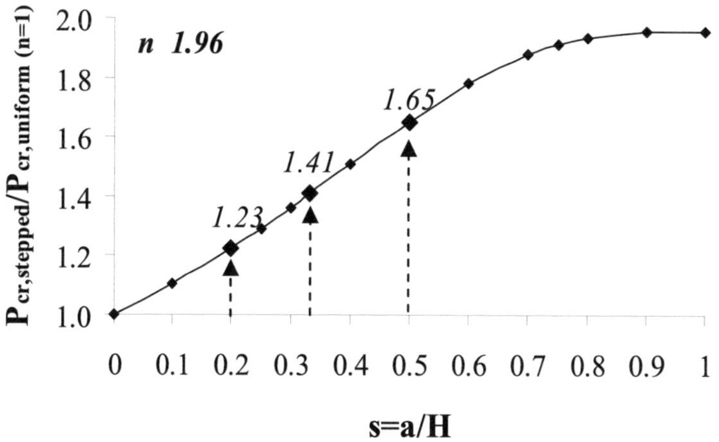

s	n=1.95794
0	1.00
0.1	1.11
0.2	1.23
0.25	1.29
0.3	1.36
0.3333	1.41
0.4	1.51
0.5	1.65
0.6	1.78
0.7	1.88
0.75	1.91
0.8	1.93
0.9	1.95
0.9999	1.96

Figure 7. Increase in critical buckling load for various stiffened length ratios (s) when stiffness ratio is $n \cong 1.96$ (VIM results)

4. Experimental studies on elastic buckling of a three-segment stepped compression member with pinned ends

The experimental part of the study is conducted in the Structures Laboratory of Civil Engineering Department in Kocaeli University. Test specimens are subjected to monotonically increasing compressive load until they buckle about their minor axis in a test setup specifically designed for such types of buckling tests (Fig. 8). Due to the height limitations of the test setup, the length of the test specimens is fixed to 2 m. To observe elastic buckling, "unstiffened" (uniform) *reference* specimens are selected to have a rather small cross section; hollow rectangular section with side dimensions of 120 mm x 40 mm and wall thickness of 4 mm, as shown in Fig. 5a. In addition to the three unstiffened specimens, named B0-1, B0-2 and B0-3, three sets of "stiffened" specimens, each of which consists of three columns with identical stiffening, are tested. To obtain comparable results, the stiffness ratio of the stiffened specimens is kept constant ($n \cong 2$) while their stiffened length ratios (s) are varied in each set. Such stiffening is attained by welding rectangular steel plates, with 100 mm width and 3 mm thickness as shown in Fig. 6a, to the wider faces of the hollow cross sections of the test specimens, in different lengths. The length of the stiffening plates is 0.4 m for the members with stiffened length ratio s=0.2, which are named B1-1, B1-2 and B1-3, approximately 0.67 m for the members with s=0.3333, named B2-1, B2-2 and B2-3, and 1.0 m for the members with s=0.5, named B3-1, B3-2 and B3-3.

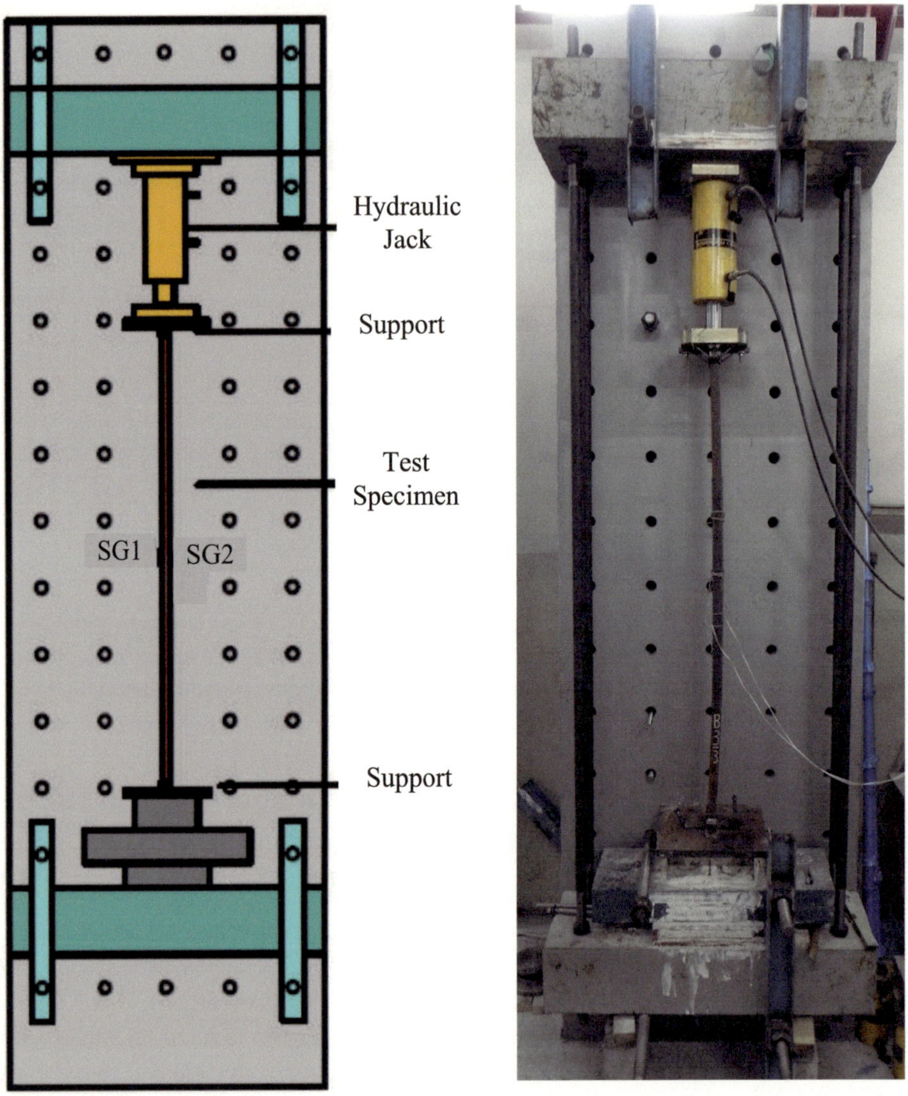

Figure 8. Test setup

As shown in Fig. 8, the test specimens are placed between the top and bottom supports in the test rig, which is rigidly connected to the strong reaction wall. To ensure minor-axis buckling of the test columns, the supports are designed in such a way that the rotation is about a single axis, resisting rotation about the orthogonal axis. In other words, the supports behave as pinned supports in minor-axis bending whereas fixed supports in major-axis bending. The compressive load is applied to the columns through a hydraulic jack placed at the top of the upper support. During the tests, in addition to the load readings, which are measured by a pressure gage, strains at the outermost fibers in the central cross section of each column are recorded via two strain gages (SG1 and SG2) (see Fig. 8).

The buckled shapes of the tested columns are presented in Fig. 9 and Fig. 10. As shown in Fig. 9a, uniform columns buckle in the shape of a half-sine wave, which is in agreement with the well-known Euler's formulation for *ideal* pinned-pinned columns. In contrast to *ideal* columns, however, test columns have *not* buckled suddenly during the tests. This is mainly due to the fact that all test specimens have unavoidable initial crookedness. Even though the amount of these imperfections remain within the tolerances specified by the specifications, they cause bending of the specimens with the initiation of loading. This is also apparent from the graphs presented in Fig. 11. These graphs plot strain gage measurements taken at the opposite sides of the column faces (SG1 and SG2) during the test of each specimen with respect to the applied load values. The divergence of strain gage readings (SG1 and SG2) from each other as the load increases clearly indicates onset of the bending under axial compression. This is compatible with the expectations since as stated by Galambos (1998), "geometric imperfections, in the form of tolerable but unavoidable out-of-straightness of the column and/or eccentricity of the axial load, will introduce bending from the onset of loading". Even though the test columns start to bend at smaller load levels, they continue to carry additional loads until they reach their "buckling" capacities, which are characterized as the peak values of their load-strain curves.

The buckling loads of all test specimens are tabulated in Table 4. When the buckling loads of three uniform columns are compared, it is observed that the buckling load for Specimen B0-3 (150.18 kN) is larger than those for Specimens B0-1 (129.60 kN) and B0-2 (128.49 kN). When Fig. 11a is examined closely, it can be observed that strain gage measurements start to deviate from each other at larger loads in Specimen B0-3 than B-01 and B0-2. Thus, it can be concluded that the capacity difference among these specimens occurs *most probably* due to the fact that the initial out-of-straightness of Specimen B0-3 is much smaller than that of B-01 and B-02. When the load-strain plots of the stiffened specimens (Fig. 11b-d) are examined, similar trends are observed for specimens with larger load values in their own sets, e.g., B2-1 and B2-3 in the third set, B3-1 in the forth set. These differences can also be attributed *partially* to the initial out-of-straightness. Unlike uniform columns, stiffened columns have additional initial imperfections due to the welding process of the stiffeners. It is now well known that welding cause unavoidable residual stresses to develop within the cross section of the member, which, in turn, can change the behavior of the member significantly. Since the columns with larger stiffened length ratios have longer welds, they are expected to have more initial imperfection. The effects of initial imperfections can also be seen from the last column of Table 4, where the ratios of experimental results to the analytical results which are obtained for *ideal* columns are presented.

For better comparison, experimental ($P_{cr,exp}$) and analytical ($P_{cr,analy}$) buckling loads are also plotted in Fig. 12. As shown in the figure, all test results lay below the analytical curve.

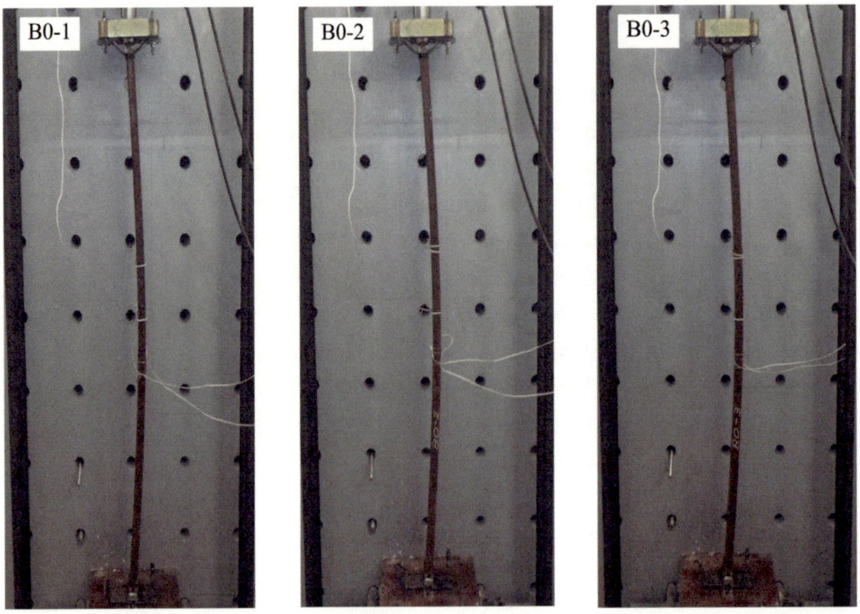

a. Unstiffened columns

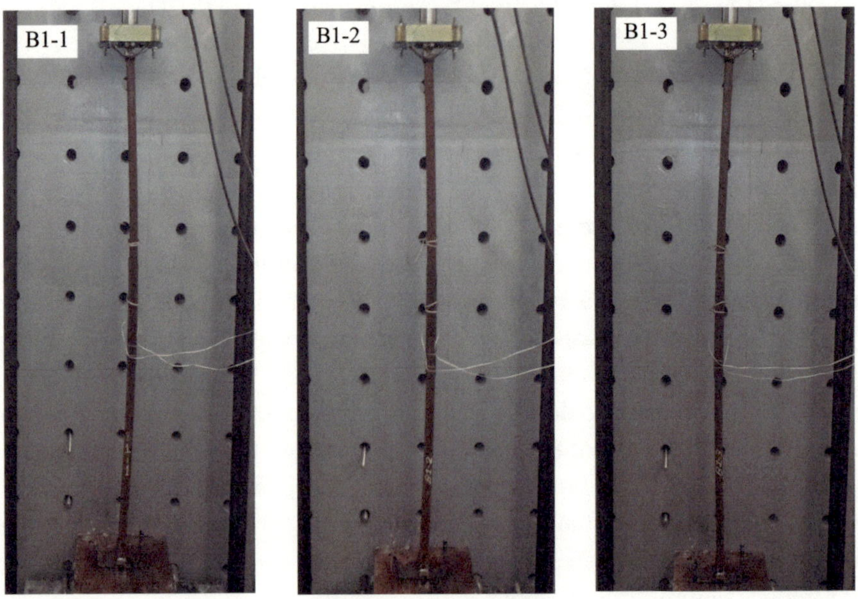

b. Stiffened columns with s=0.2

Figure 9. Buckled shapes of unstiffened and stiffened (with s=0.2) test specimens

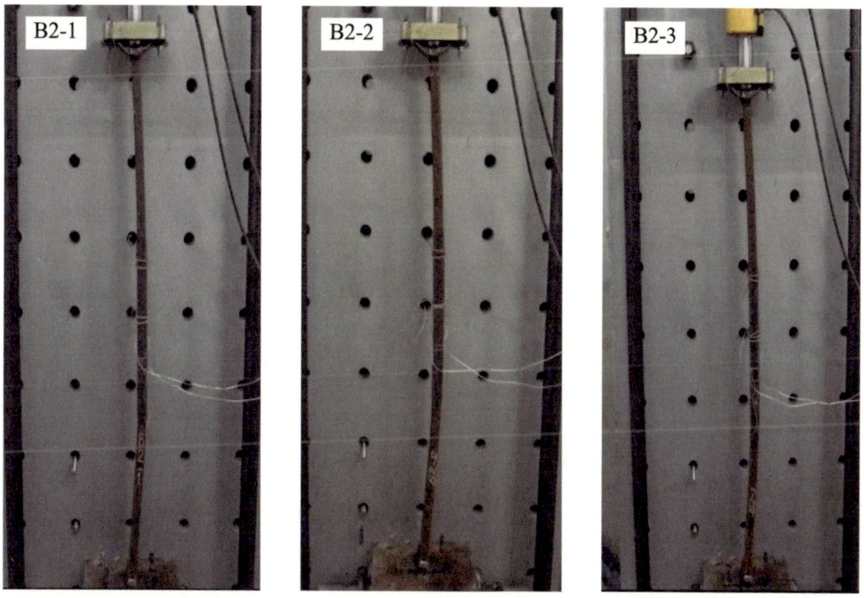

a. Stiffened columns with s=0.3333

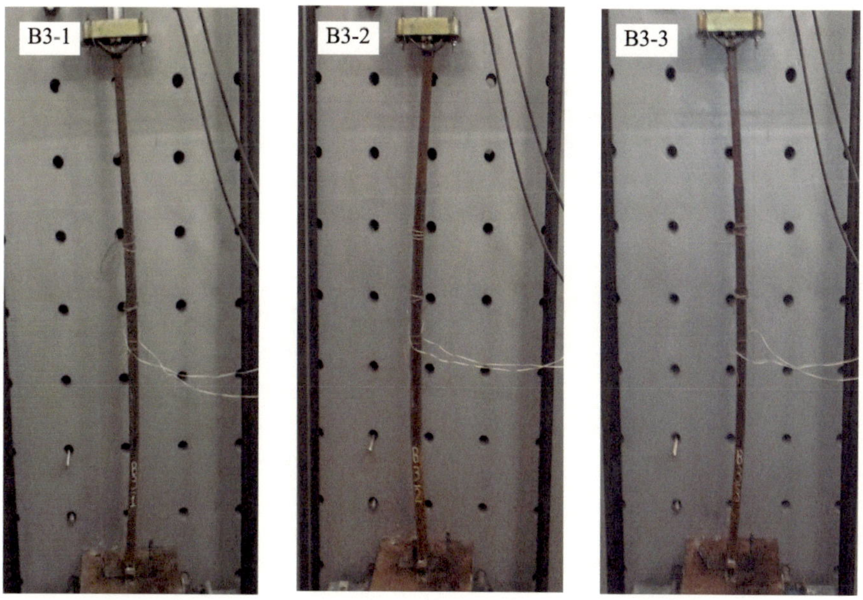

b. Stiffened columns with s=0.5

Figure 10. Buckled shapes of stiffened test specimens with s=0.3333 and s=0.5

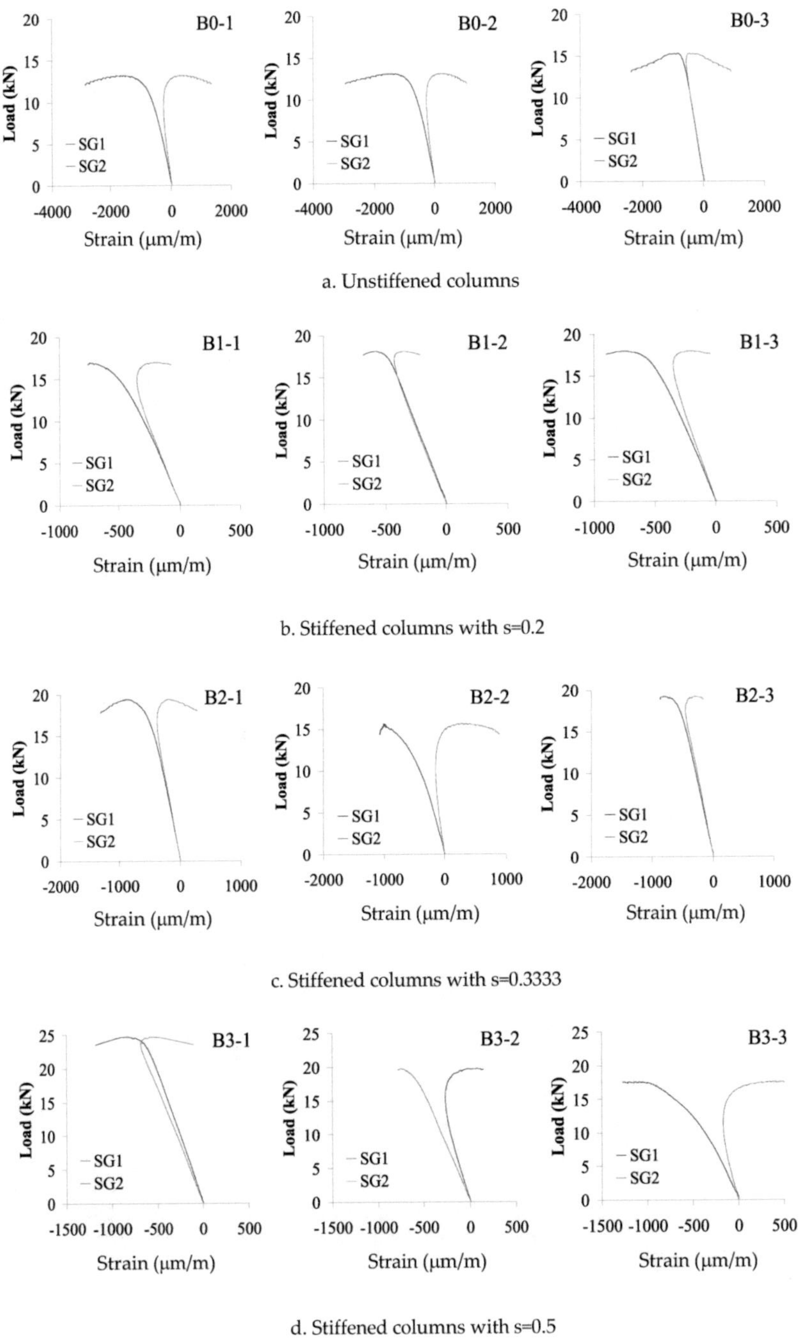

Figure 11. Load versus strain gage measurements for the test specimens

Specimen	s	$P_{cr,exp}$ (kN)	$P_{cr,analy}$ (kN)	$P_{cr,exp}$ / $P_{cr,analy}$
B0-1	0	129.60	157.42	0.823
B0-2		128.49		0.816
B0-3		150.18		0.954
B1-1	0.2	166.31	192.98	0.862
B1-2		177.44		0.919
B1-3		176.32		0.914
B2-1	0.3333	190.23	221.78	0.858
B2-2		153.52		0.692
B2-3		188.56		0.850
B3-1	0.5	241.96	260.10	0.930
B3-2		194.12		0.746
B3-3		172.43		0.663

Table 4. Experimental buckling loads for uniform and stiffened columns compared with the analytical predictions

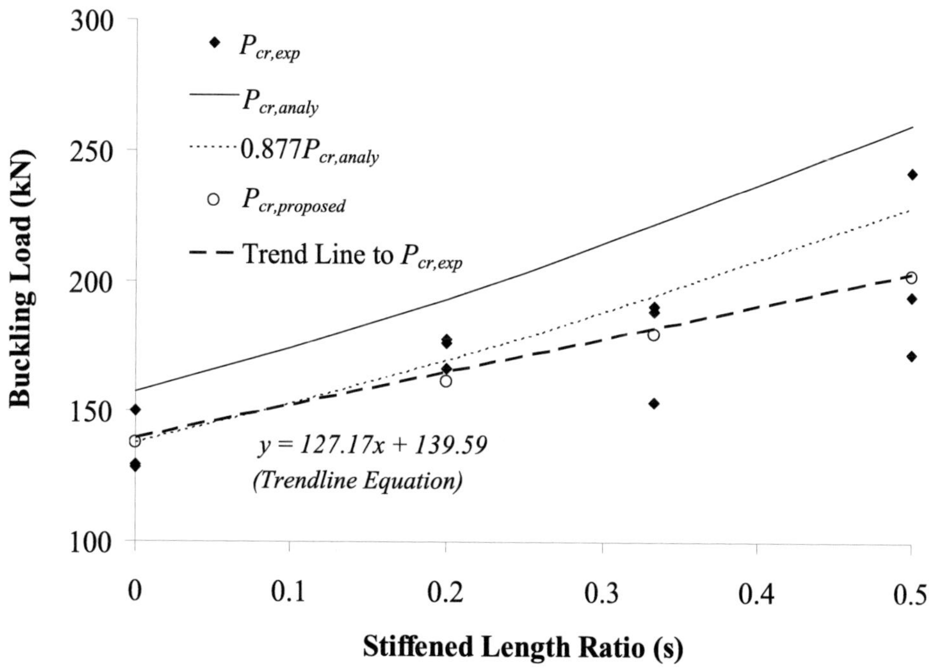

Figure 12. Experimental results compared with analytical and modified analytical buckling loads

It is important to note that most design specifications modify the buckling load equations derived for *ideal* columns to take into account the effects of initial out-of-straightness of the columns in the design of compression members. As an example, to reflect an initial out-of-straightness of about 1/1500, AISC (2010) modifies the "Euler" load by multiplying with a factor of 0.877 in the calculation of compressive capacity of elastically buckling members

(Salmon et al., 2009). By applying a similar modification to the analytical results obtained in this study for *ideal* three-segment compression members, a *more realistic* analytical curve is drawn. This curve is plotted in Fig. 12 with a label '0.877 $P_{cr,analy}$'. From Fig. 12, it is seen that the "modified" analytical curve *almost* "averages" most of the test results. The larger discrepancies observed in stiffened specimens with s=0.3333 and s=0.5 are believed to be resulted from the residual stresses locked in the specimens during welding of the steel stiffening plates, which highly depends on quality of workmanship. For this reason, while calculating the buckling load of a multi-segment compression member formed by welding, not only the initial out-of-straightness of the member, but also the effects of welding have to be taken into account. Considering that stiffened columns will always have more initial imperfections than uniform columns, it is suggested that a smaller modification factor be used in the design of multi-segment columns. Based on the limited test data obtained in the experimental phase of this study, the following modification factor is proposed to be used in the design of three-segment symmetric steel compression members formed by welding steel stiffening plates:

$$MF = \left(0.877 - 0.2s\right) \tag{25}$$

where s is the stiffened length ratio of the compression member, which equals to the weld length in the stiffened members. Thus, the proposed buckling load ($P_{cr,proposed}$) for such a member can be computed by modifying the analytical buckling load ($P_{cr,analy}$) as in the following expression:

$$P_{cr,proposed} = MF \times P_{cr,analy} \tag{26}$$

The proposed buckling loads for the multi-segment columns tested in the experimental part of this study are computed using Eq. (26) with Eq. (25) and plotted in Fig. 12 with a label '$P_{cr,proposed}$'. For easier comparison, a linear trend line fitted to the experimental data is also plotted in the same figure. Fig. 12 shows perfect match of design values of buckling loads with the trend line. While using Eq. (25), it should be kept in mind that the modification factor proposed in this paper is derived based on the limited test data obtained in the experimental part of this study and needs being verified by further studies.

5. Conclusion

In an attempt to design economic and aesthetic structures, many engineers nowadays prefer to use nonuniform members in their designs. Strengthening a steel braced structure which have insufficient lateral resistant by stiffening the braces through welding additional steel plates or wrapping fiber reinforced polymers in partial length is, for example, a special application of use of multi-segment nonuniform members in earthquake resistant structural engineering. The stability analysis of multi-segment (stepped) members is usually very complicated, however, due to the complex differential equations to be solved. In fact, most of the design formulae/charts given in design specifications are developed for uniform members. For this reason, there is a need for a practical tool to analyze buckling behavior of nonuniform members.

In this study, elastic buckling behavior of three-segment symmetric stepped compression members with pinned ends is analyzed using three different approaches: (i) analytical, (ii) numerical and (iii) experimental approaches. In the analytical study, first the governing equations of the studied stability problem are derived. Then, exact solution is obtained. Since exact solution requires finding the smallest root of a rather complex characteristic equation which highly depends on initial guess, the governing equations are also solved using a recently developed analytical technique, called Variational Iteration Method (VIM), and it is shown that it is much easier to solve the characteristic equation derived using VIM. The problem is also handled, for some special cases, by using widely known structural analysis program SAP2000 (CSI, 2008). Agreement of numerical results with analytical results indicates that such an analysis program can also be effectively used in stability analysis of stepped columns. Finally, aiming at the verification of the analytical results, the buckling loads of steel columns with hollow rectangular cross section stiffened, in partial length, by welding steel plates are investigated experimentally. Experimental results point out that the buckling loads obtained for *ideal* columns using analytical formulations have to be modified to reflect the initial imperfections. If welding is used while forming the stiffened members, as done in this study, not only the initial out-of-straightness, but also the effects of welding have to be considered in this modification. Based on the limited test data, a modification factor which is a linear function of the stiffened length ratio is proposed for three-segment symmetric steel compression members formed by welding steel plates in the stiffened regions.

Author details

Seval Pinarbasi Cuhadaroglu, Erkan Akpinar,
Fuad Okay, Hilal Meydanli Atalay and Sevket Ozden
Kocaeli University, Turkey

6. References

Abulwafa, E.M.; Abdou, M.A. & Mahmoud, A.A. (2007). Nonlinear fluid flows in pipe-like domain problem using variational iteration method. *Chaos Solitons & Fractals*, Vol.32, No.4, pp. 1384–1397.

American Institute of Steel Construction (AISC). (2010). *Specification for Structural Steel Buildings (AISC 360-10)*, Chicago.

Atay, M.T. & Coskun, S.B. (2009). Elastic stability of Euler columns with a continuous elastic restraint using variational iteration method. *Computers and Mathematics with Applications*, Vol.58, pp. 2528-2534.

Batiha, B.; Noorani, M.S.M. & Hashim, I. (2007). Application of variational iteration method to heat- and wave-like equations. *Physics Letters A*, Vol. 369, pp. 55-61.

Computers and Structures Inc. (CSI) (2008) *SAP2000 Static and Dynamic Finite Element Analysis of Structures* (Advanced 12.0.0), Berkeley, California.

Coskun, S.B. & Atay, M.T. (2007). Analysis of convective straight and radial fins with temperature- dependent thermal conductivity using variational iteration method with comparison with respect to finite element analysis. *Mathematical Problems in Engineering*, Article ID: 42072.

Coskun, S.B. & Atay, M.T. (2008). Fin efficiency analysis of convective straight fins with temperature dependent thermal conductivity using variational iteration method. *Applied Thermal Engineering*, Vol.28, No.17-18, pp. 2345-2352.

Coskun, S.B. & Atay, M.T. (2009). Determination of critical buckling load for elastic columns of constant and variable cross-sections using variational iteration method. *Computers and Mathematics with Applications*, Vol.58, pp. 2260-2266.

Ganji, D.D. & Sadighi, A. (2007). Application of homotopy-perturbation and variational iteration methods to nonlinear heat transfer and porous media equations. *Journal of Computational and Applied Mathematics*, Vol.207, pp. 24-34.

Galambos, T.V. (1998). *Guide to Stability Design Criteria for Metal Structures* (fifth edition), John Wiley & Sons, Inc., ISBN 0-471-12742-6, NewYork.

He, J.H. (1999). Variational iteration method - a kind of nonlinear analytical technique: some examples. *International Journal of Non Linear Mechanics*, Vol.34, No.4, pp. 699-708.

He, J.H.; Wu, G.C. & Austin, F. (2010). The variational iteration method which should be followed. *Nonlinear Science Letters A*, Vol.1, No.1, pp. 1-30.

Li, Q.S. (2001). Buckling of multi-step non-uniform beams with elastically restrained boundary conditions. *Journal of Constructional Steel Research*, Vol.57, pp. 753–777.

Miansari, M.; Ganji, D.D. & Miansari M. (2008). Application of He's variational iteration method to nonlinear heat transfer equations. *Physics Letters A*, Vol. 372, pp. 779-785.

Okay, F.; Atay, M.T. & Coskun S.B. (2010). Determination of buckling loads and mode shapes of a heavy vertical column under its own weight using the variational iteration method. *International Journal of Nonlinear Science Numerical Simulation*, Vol.11, No.10, pp. 851-857.

Ozturk, B. (2009). Free vibration analysis of beam on elastic foundation by variational iteration method. *International Journal of Non Linear Mechanics*, Vol.10, No.10, pp. 1255-1262.

Pinarbasi, S. (2011). Lateral torsional buckling of rectangular beams using variational iteration method. *Scientific Research and Essays*, Vol.6, No.6, pp. 1445-1457.

Salmon, C.G.; Johnson, E.J. & Malhas, F.A. (2009). *Steel Structures, Design and Behavior* (fifth edition), Pearson, Prentice Hall, ISBN-10: 0-13-188556-1, New Jersey.

Sweilan, N.H. & Khader, M.M. (2007). Variational iteration method for one dimensional nonlinear thermoelasticity. *Chaos Solitons & Fractals*, Vol.32, No.1, pp. 145-149.

Timoshenko, S.P. & Gere, J.M. (1961). *Theory of Elastic Stability* (second edition), McGraw-Hill Book Company, ISBN- 0-07-085821-7, New York.

Chapter 6

Elastic Stability Analysis of Euler Columns Using Analytical Approximate Techniques

Safa Bozkurt Coşkun and Baki Öztürk

Additional information is available at the end of the chapter

http://dx.doi.org/10.5772/45940

1. Introduction

In most of the real world engineering applications, stability analysis of compressed members is very crucial. There have been many researches dedicated to the buckling behavior of axially compressed members. On the other hand, obtaining analytical solutions for the buckling behavior of columns with variable cross-section subjected to complicated load configurations are almost impossible in most of the cases. Some of the works related to obtaining analytical or analytical approximate solutions for the column buckling problem are provided below.

The problems of buckling of columns under variable distributed axial loads were solved in detail by Vaziri and Xie [1] and others. Some analytical closed-form solutions are given by Dinnik [2], Karman and Biot [3], Morley[4], Timoshenko and Gere [5] and others. One of the detailed references related to the structural stability topic is written by Simitses and Hodges [6] with detailed discussions. Iyengar [7] made some analysis on buckling of uniform with several elastic supports. Wang et al. [8] have given exact mathematical solutions for buckling of structural members for various cases of columns, beams, arches, rings, plates and shells. Ermopoulos [9] found the solution for buckling of tapered bars axially compressed by concentrated loads applied at various locations along their axes. Li [10] gave the exact solution for buckling of non-uniform columns under axially concentrated and distributed loading. Lee and Kuo [11] established an analytical procedure to investigate the elastic stability of a column with elastic supports at the ends under uniformly distributed follower forces. Furthermore, Gere and Carter [12] investigated and established the exact analytical solutions for buckling of several special types of tapered columns with simple boundary conditions. Solution of the problem of buckling of elastic columns with step varying thickness is established by Arbabei and Li [13]. Stability problems of a uniform bar with several elastic supports using the moment-distribution method were analyzed by Kerekes [14]. The research of Siginer [15] was about the stability of a column whose flexural stiffness has a continuous linear variation along the column. Moreover, the analytical solutions of a multi-step bar with varying cross section were obtained by Li et al. [16-18]. The energy method was used by Sampaio et al. [19] to find the solution for the problem of buckling behavior of inclined beam-column. Some of the important researchers who studied the mechanical behavior of beam-columns are Keller [20], Tadjbakhsh and Keller [21] and Taylor [22]. Later on, analytical approximate techniques were used for the stability analysis of elastic columns. Coşkun and Atay [23] and Atay and Coşkun [24] studied column buckling problems for the columns with variable flexural stiffness and for the columns with continuous elastic restraints by using the variational iteration method which produces analytical approximations. Coşkun [25, 26] used the homotopy perturbation method for buckling of Euler columns on elastic foundations and tilt-buckling of variable stiffness columns. Pınarbaşı [27] also analyzed the stability of nonuniform rectangular beams using homotopy perturbation method. These techniques were also used successfully in the vibration analysis of Euler-Bernoulli beams and in the vibration of beams on elastic foundations. [28-29]

Recently, by the emergence of new and innovative semi analytical approximation methods, research on this subject has gained momentum. Analytical approximate solution techniques are used widely to solve nonlinear ordinary or partial differential equations, integrodifferential equations, delay equations, etc. The main advantage of employing such techniques is that the problems are considered in a more realistic manner, and the solution obtained is a continuous function which is not the case for the solutions obtained by discretized solution techniques.

The methods that will be used throughout this study are, Adomian Decomposition Method (ADM), Variational Iteration Method (VIM) and Homotopy Perturbation Method (HPM). Each technique will be explained first, and then all will be applied to a selected case study related to the topic of the article.

2. Problem formulation

Derivation of governing equations related to stability analysis is given in detail in Timoshenko and Gere [5], Simitses and Hodges [6], and Wang et al. [8]. The reader can also refer to any textbook related to the subject. In this section, only the governing equation will be given for the related cases.

Consider the elastic columns given in Fig.1. The governing equation for the buckling of such columns is

$$\frac{d^2y}{dx^2}\left[EI(x)\frac{d^2y}{dx^2}\right]+P\frac{d^2y}{dx^2}=0 \tag{1}$$

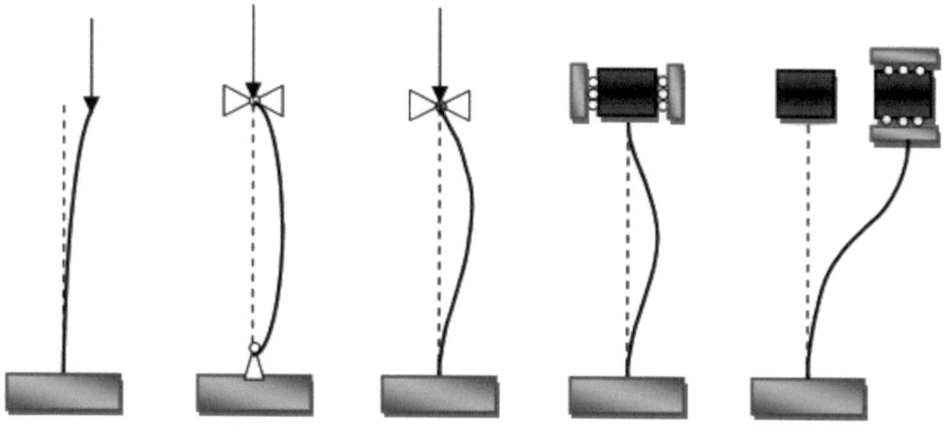

Figure 1. Elastic columns with various end conditions

In the case of constant flexural rigidity (*i.e. EI* is constant), Eq.(1) becomes

$$\frac{d^4y}{dx^4}+\frac{P}{EI}\frac{d^2y}{dx^2}=0 \tag{2}$$

where EI is the flexural rigidity of the column, and P is the applied load. Both Eqs. (2) and (3) are solved due to end conditions of the column. Some of these conditions are shown in Fig.1. In this figure, letters are used for a simplification to describe the support conditions of the column. The first letter stands for the support at the bottom and the second letter for the top. Hence, CF is *Clamped-Fixed,* PP is *Pinned-Pinned,* C-P is *Clamped-Pinned* and C-S is *Clamped-Sliding Restraint.*

The governing equations (1) and (2) are both solved with respect to the problem's end conditions. The end conditions for the columns shown in Fig.1 are given below:

Pin support:

$$y=0 \text{ and } \frac{d^2y}{dx^2}=0 \tag{3}$$

Clamped support:

$$y=0 \text{ and } \frac{dy}{dx}=0 \tag{4}$$

Free end:

$$\text{and } \frac{d^3y}{dx^3}+\frac{P}{EI}\frac{dy}{dx}=0 \tag{5}$$

Sliding restraint:

$$\frac{dy}{dx} = 0 \text{ and} \tag{6}$$

The governing equation given in Eq.(2) is a fourth order differential equation with constant coefficients which makes it possible to obtain analytical solutions easily. However, Eq.(1) includes variable coefficients due to variable flexural rigidity. For this type of differential equations, analytical solutions are limited for the special cases of *EI(x)* only. It is not possible to obtain a solution for any form of the function *EI(x)*.

In some problems, obtaining analytical solutions is very difficult even for a constant coefficient governing equation. Consider the buckling of a column on an elastic foundation shown in Fig.2.

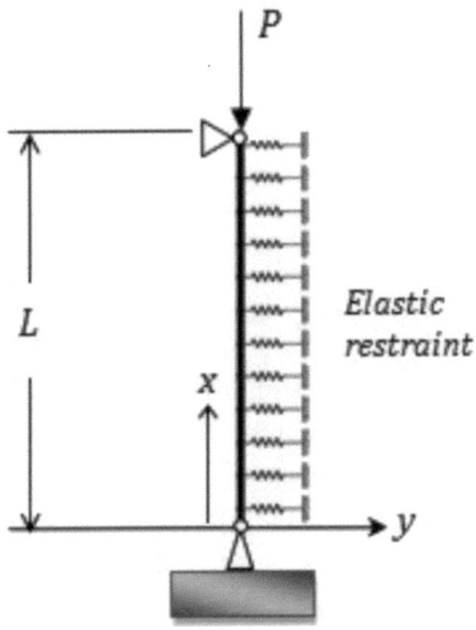

Figure 2. Column with continuous elastic restraints.

The governing equation for the column in Fig.2 is

$$\frac{d^2y}{dx^2}\left[EI(x)\frac{d^2y}{dx^2}\right] + P\frac{d^2y}{dx^2} + ky = 0 \tag{7}$$

which, for the constant *EI* becomes

$$\frac{d^4y}{dx^4} + \frac{P}{EI}\frac{d^2y}{dx^2} + \frac{k}{EI}y = 0 \tag{8}$$

In Eqs.(7) and (8), k is the stiffness parameter for the elastic restraint. The solution of Eq.(8) for the *CF* column is given in [8] as

$$\left[\alpha(S^2+T^2)-2S^2T^2\right]\cos T\cos S-\alpha(S^2+T^2)+(S^4+T^4)+ \\ ST\left[2\alpha-(S^2+T^2)\right]\sin T\sin S=0 \tag{9}$$

Although Eq.(8) is a linear equation with constant coefficients, obtaining a solution from Eq.(9) is not that easy. It is very interesting that, even with a software, one can not easily produce the buckling loads in a sequential order from Eq.(9). In view of this experience, an analytical solution for Eq.(7) is almost impossible to obtain except very limited *EI(x)* choices.

Hence, analytical approximate techniques are efficient alternatives for solving these problems. By the use of these techniques, a solution which is continuous in the problem domain is possible for any variation in flexural rigidity. These techniques produce the buckling loads in a sequential order, and it is also very easy to obtain the buckling mode shapes from the solution provided by the method used. These are great advantages in the solution of such problems.

3. The methods used in the elastic stability analysis of Euler columns

3.1. Adomian Decomposition Method (ADM)

In the ADM a differential equation of the following form is considered

$$Lu+Ru+Nu=g(x) \tag{10}$$

where, L is the linear operator which is highest order derivative, R is the remainder of linear operator including derivatives of less order than L, Nu represents the nonlinear terms, and g is the source term. Eq.(10) can be rearranged as

$$Lu=g(x)-Ru-Nu \tag{11}$$

Applying the inverse operator L^{-1} to both sides of Eq.(11) and employing given conditions; we obtain

$$u=L^{-1}\{g(x)\}-L^{-1}(Ru)-L^{-1}(Nu) \tag{12}$$

After integrating source term and combining it with the terms arising from given conditions of the problem, a function $f(x)$ is defined in the equation as

$$u=f(x)-L^{-1}(Ru)-L^{-1}(Nu) \tag{13}$$

The nonlinear operator $Nu=F(u)$ is represented by an infinite series of specially generated (Adomian) polynomials for the specific nonlinearity. Assuming Nu is analytical, we write

$$F(u)=\sum_{k=0}^{\infty}A_k \tag{14}$$

The polynomials A_k's are generated for all kinds of nonlinearity, so that they depend only on u_o to u_k components and can be produced by the following algorithm.

$$A_0 = F(u_0) \tag{15}$$

$$A_1 = u_1 F'(u_0) \tag{16}$$

$$A_2 = u_2 F'(u_0) + \frac{1}{2!} u_1^2 F''(u_0) \tag{17}$$

$$A_3 = u_3 F'(u_0) + u_1 u_2 F''(u_0) + \frac{1}{3!} u_1^3 F'''(u_0) \tag{18}$$

$$\vdots$$

The reader can refer to [30, 31] for the algorithms used in formulating Adomian polynomials. The solution $u(x)$ is defined by the following series

$$u = \sum_{k=0}^{\infty} u_k \tag{19}$$

where, the components of the series are determined recursively as follows:

$$u_0 = f(x) \tag{20}$$

$$u_{k+1} = -L^{-1}\left(Ru_k\right) - L^{-1}\left(A_k\right), \quad k \geq 0 \tag{21}$$

3.2. Variational Iteration Method (VIM)

According to VIM, the following differential equation may be considered:

$$Lu + Nu = g(x) \tag{22}$$

where L is a linear operator, N is a nonlinear operator, and $g(x)$ is an inhomogeneous source term. Based on VIM, a correct functional can be constructed as follows:

$$u_{n+1} = u_n + \int_0^x \lambda(\xi)\left\{Lu_n(\xi) + N\tilde{u}_n(\xi) - g(\xi)\right\} d\xi \tag{23}$$

where λ is a general Lagrangian multiplier which can be identified optimally via the variational theory. The subscript n denotes the nth-order approximation, $\tilde{u}$ is considered as a restricted variation *i.e.* $\delta\tilde{u} = 0$. By solving the differential equation for λ obtained from Eq.(23) in view of $\delta\tilde{u} = 0$ with respect to its boundary conditions, Lagrangian multiplier $\lambda(\xi)$ can be obtained. For further details of the method the reader can refer to [32].

3.3. Homotopy Perturbation Method (HPM)

HPM provides an analytical approximate solution for problems at hand as the other previously explained techniques. Brief theoretical steps for the equation of following type can be given as

$$L(u) + N(u) = f(r) \ , \ r \in \Omega \tag{24}$$

with boundary conditions $B(u, \partial u/\partial n) = 0$. In Eq.(24) L is a linear operator, N is a nonlinear operator, B is a boundary operator, and $f(r)$ is a known analytic function. HPM defines homotopy as

$$v(r,p) = \Omega \times [0,1] \to R \tag{25}$$

which satisfies the following inequalities:

$$H(v,p) = (1-p)[L(v) - L(u_0)] + p[L(v) + N(v) - f(r)] = 0 \tag{26}$$

or

$$H(v,p) = L(v) - L(u_0) + pL(u_0) + p[N(v) - f(r)] = 0 \tag{27}$$

where $r \in \Omega$ and $p \in [0,1]$ is an imbedding parameter, u_0 is an initial approximation which satisfies the boundary conditions. Obviously, from Eq.(26) and Eq.(27) , we have :

$$H(v,0) = L(v) - L(u_0) = 0 \tag{28}$$

$$H(v,1) = L(v) + N(v) - f(r) = 0 \tag{29}$$

As p is changing from zero to unity, so is that of $v(r,p)$ from u_0 to $u(r)$. In topology, this deformation $L(v) - L(u_0)$ and $L(v) + N(v) - f(r)$ are called homotopic. The basic assumption is that the solutions of Eq.(34) and Eq.(35) can be expressed as a power series in p such that:

$$v = v_0 + pv_1 + p^2 v_2 + p^3 v_3 + \dots \tag{30}$$

The approximate solution of $L(u) + N(u) = f(r) \ , \ r \in \Omega$ can be obtained as :

$$u = \lim_{p \to 1} v = v_0 + v_1 + v_2 + v_3 + \dots \tag{31}$$

The convergence of the series in Eq.(31) has been proved in [33]. The method is described in detail in references [33-36].

4. Case studies

4.1. Buckling of a clamped-pinned column

The governing equation for this case was previously given in Eq.(1). ADM, VIM and HPM will be applied to this equation in order to compute the buckling loads for the clamped-pinned column with constant flexural stiffness, *i.e.* constant *EI*, and variable flexural stiffness, *i.e.* variable *EI* with its corresponding mode shapes. To achieve this aim, a circular rod is defined with an exponentially varying radius. The case is given in Fig.3, and rod and its associated boundary conditions are also provided in Eqs.(3-4). As a case study, first the formulations for constant stiffness column by using ADM, VIM and HPM are given, and then applied to the governing equation of the problem. Afterwards, a variable flexural rigidity will be defined for the same column, and the same techniques will be used for the analysis.

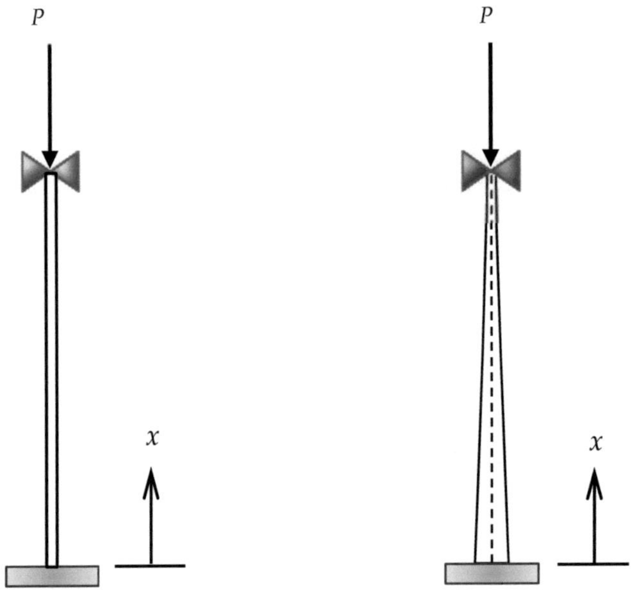

Figure 3. *CP* column with constant and variable flexural rigidity

4.2. Formulation of the algorithms for uniform column

4.2.1. ADM

The linear operator and its inverse operator for Eq.(2) is

$$L(\cdot) = \frac{d^4}{dx^4}(\cdot) \tag{32}$$

$$L^{-1}(\cdot) = \int_0^x\int_0^x\int_0^x\int_0^x (\cdot)\, dx\, dx\, dx\, dx \tag{33}$$

To keep the formulation a general one for all configurations to be considered, the boundary conditions are chosen as $Y(0) = A$, $Y'(0) = B$, $Y''(0) = C$ and $Y'''(0) = D$. Suitable values should be replaced in the formulation with these constants. In this case, $A = 0$ and $B = 0$ should be inserted for the *CP* column. Hence, the equation to be solved and the recursive algorithm can be given as

$$LY = \zeta Y'' \tag{34}$$

$$Y = A + Bx + C\frac{x^2}{2!} + D\frac{x^3}{3!} + L^{-1}(\zeta Y'') \tag{35}$$

$$Y_{n+1} = L^{-1}(\zeta Y_n''), \quad n \geq 0 \tag{36}$$

Finally, the solution is defined by

$$Y = Y_0 + Y_1 + Y_2 + Y_3 + ... \tag{37}$$

4.2.2. VIM

Based on the formulation given previously, Lagrange multiplier, λ would be obtained for the governing equation, *i.e.* Eq.(2), as

$$\lambda(\xi) = \frac{(\xi - x)^3}{3!} \tag{38}$$

An iterative algorithm can be constructed inserting Lagrange multiplier and governing equation into the formulation given in Eq.(31) as

$$Y_{n+1} = Y_n + \int_0^x \lambda(\xi)\left\{Y_n^{iv}(\xi) + \zeta \tilde{Y}_n''(\xi)\right\} d\xi \tag{39}$$

where ζ is normalized buckling load for the column considered. Initial approximation for the algorithm is chosen as the solution of $LY = 0$ which is a cubic polynomial with four unknowns to be determined by the end conditions of the column.

4.2.3. HPM

Based on the formulation, Eq.(2) can be divided into two parts as

$$LY = Y^{iv} \tag{40}$$

$$NY = \zeta Y'' \tag{41}$$

The solution can be expressed as a power series in p such that

$$Y = Y_0 + pY_1 + p^2Y_2 + p^3Y_3 + ... \tag{42}$$

Inserting Eq.(50) into Eq.(35) provides a solution algorithm as

$$Y_0^{iv} - y_0^{iv} = 0 \tag{43}$$

$$Y_1^{iv} + y_0^{iv} + \zeta^4 Y_0'' = 0 \tag{44}$$

$$Y_n^{iv} + \zeta Y_{n-1}'' = 0, \quad n \geq 2 \tag{45}$$

Hence, an approximate solution would be obtained as

$$Y = Y_0 + Y_1 + Y_2 + Y_3 + \dots \tag{46}$$

Initial guess is very important for the convergence of solution in HPM. A cubic polynomial with four unknown coefficients can be chosen as an initial guess which was shown previously to be an effective one in problems related to Euler beams and columns [23-29].

4.3. Computation of buckling loads

By the use of described techniques, an iterative procedure is constructed and a polynomial including the unknown coefficients resulting from the initial guess is produced as the solution to the governing equation. Besides two unknowns from the initial guess, an additional unknown ζ also exists in the solution. Applying far end boundary conditions to the solution produces a linear algebraic system of equations which can be defined in a matrix form as

$$\left[M(\zeta)\right]\{\alpha\} = \{0\} \tag{47}$$

where $\{\alpha\} = \langle A, B \rangle^T$. For a nontrivial solution, determinant of coefficient matrix must be zero. Determinant of matrix $\left[M(\zeta)\right]$ yields a characteristic equation in terms of ζ. Positive real roots of this equation are the normalized buckling loads for the Clamped-Pinned column.

4.4. Determination of buckling mode shapes

Buckling mode shapes for the column can also be obtained from the polynomial approximations by the methods considered in this study. Introducing, the buckling loads into the solution, normalized polynomial eigen functions for the mode shapes are obtained from

$$\overline{Y}_j = \frac{Y_N\left(x, \zeta_j\right)}{\left[\int_0^1 \left|Y_N\left(x, \zeta_j\right)\right|^2 dx\right]^{1/2}}, \quad j = 1, 2, 3, \dots \tag{48}$$

The same approach can also be employed to predict mode shapes for the cases including variable flexural stiffness.

4.5. Analysis of a uniform column

After applying the procedures explained in the text, the following results are obtained for the buckling loads. Comparison with the exact solutions is also provided in order that one can observe an excellent agreement between the exact results and the computed results.

Twenty iterations are conducted for each method, and the computed values are compared with the corresponding exact values for the first four modes of buckling in the following table.

Mode	*Exact*	*ADM*	*VIM*	*HPM*
1	20.19072856	20.19072856	20.19072856	20.19072856
2	59.67951594	59.67951594	59.67951594	59.67951594
3	118.89986916	118.89986857	118.89986857	118.89986868
4	197.85781119	197.88525697	197.88522951	197.88520511

Table 1. Comparison of normalized buckling loads (PL^2 / EI) for the *CP* column

From the table it can be seen that the computed values are highly accurate which show that the techniques used in the analysis are very effective. Only a few iterations are enough to obtain the critical buckling load which is Mode 1. Additional modes require additional iterations. The table also shows that additional two or three iterations will produce an excellent agreement for Mode 4. Even with twenty iterations, the error is less than 0.014% for all the methods used in the analyses.

The buckling mode shapes of uniform column for the first four modes are depicted in Fig.4. To prevent a possible confusion to the reader, the exact mode shapes and the computed ones are not shown separately in the figure since the obtained mode shapes coincide with the exact ones.

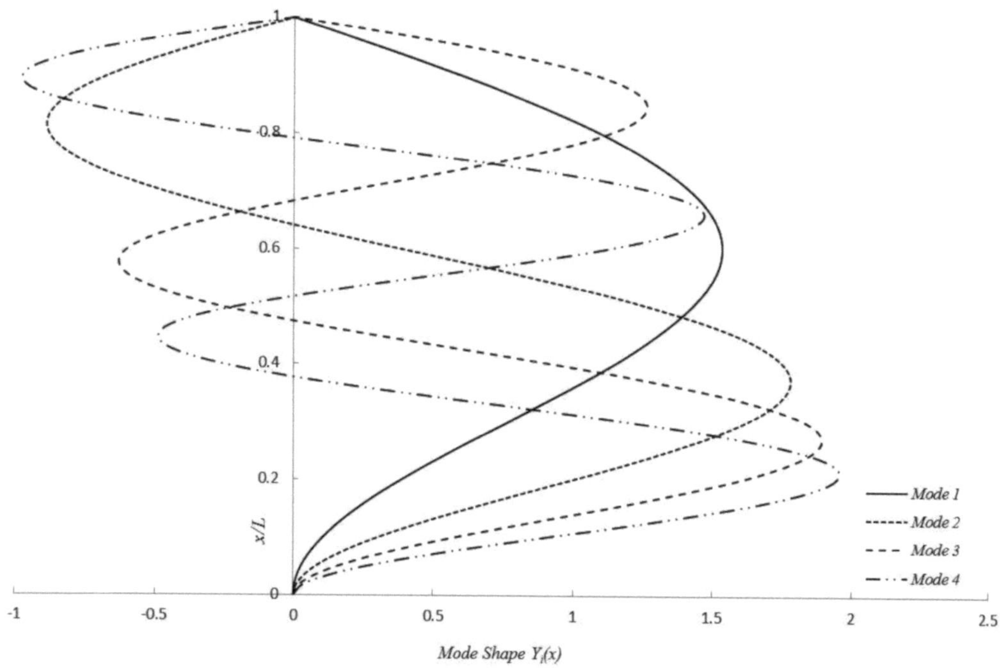

Figure 4. Buckling modes of *CP* column.

4.6. Buckling of a rod with variable cross-section

A circular rod having a radius changing exponentially is considered in this case. Such a rod is shown below in Fig.5. The function representing the radius would be as

$$R(x) = R_0\,\mathrm{e}^{-ax} \tag{49}$$

where R_o is the radius at the bottom end, L is the length of the rod and $aL \leq 1$.

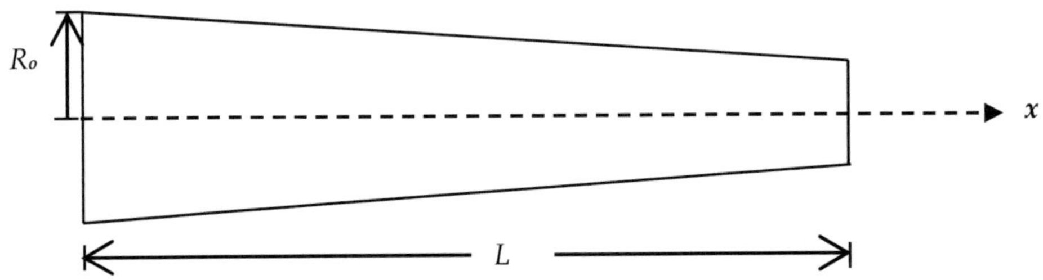

Figure 5. Circular rod with a variable cross-section

Employing Eq.(49), cross-sectional area and moment of inertia for a section at an arbitrary point x becomes:

$$A(x) = A_0\,\mathrm{e}^{-2ax} \tag{50}$$

$$I(x) = I_0\,\mathrm{e}^{-4ax} \tag{51}$$

where

$$A_0 = \pi R_0^2 \tag{52}$$

$$I_0 = \frac{\pi R_0^4}{4} \tag{53}$$

Governing equation for the rod was previously given in Eq.(1) as

$$\frac{d^2y}{dx^2}\left[EI(x)\frac{d^2y}{dx^2}\right] + P\frac{d^2y}{dx^2} = 0$$

4.6.1. *Formulation of the algorithms*

4.6.1.1. ADM

Application of ADM leads to the following

$$Y^{iv} - 8aY''' + \left(16a^2 + \zeta\right)\psi(x)Y'' = 0 \tag{54}$$

where

$$\psi(x) = e^{4ax} \tag{55}$$

and, where ζ is normalized buckling load PL^2 / EI_0. Once ζ is provided by ADM, buckling mode shapes for the rod can also be easily produced from the solution.

ADM gives the following formulation with the previously defined fourth order linear operator.

$$Y = A\frac{x^2}{2!} + B\frac{x^3}{3!} + L^{-1}\left(8aY''' - \left(16a^2 + \zeta\right)\psi(x)Y''\right) \tag{56}$$

4.6.1.2. VIM

Lagrange multiplier is the same as used in the uniform column case due to the fourth order derivative in Eq.(38). Hence, an algorithm by using VIM can be constructed as

$$Y_{n+1} = Y_n + \int_0^x \lambda(\xi)\left\{Y_n^{iv} - 8a\tilde{Y}_n''' + \left(16a^2 + \zeta\right)\psi(x)\tilde{Y}_n''\right\} d\xi \tag{57}$$

4.6.1.3. HPM

Application of HPM produces the following set of recursive equations as the solution algorithm.

$$Y_0^{iv} - y_0^{iv} = 0 \tag{58}$$

$$Y_1^{iv} + y_0^{iv} - 8aY_0''' + \left(16a^2 + \zeta\right)\psi(x)Y_0'' = 0 \tag{59}$$

$$Y_n - 8aY_{n-1}''' + \left(16a^2 + \zeta\right)\psi(x)Y_{n-1}'' = 0, \quad n \geq 2 \tag{60}$$

4.6.2. Results of the analyses

The proposed formulations are applied for two different variations, *i.e.* $aL = 0.1$ and $aL = 0.2$. Twenty iterations are conducted for each method, and the computed normalized buckling load PL^2 / EI_0 values are given for the first four modes of buckling in Tables 2 and 3.

Mode	*ADM*	*VIM*	*HPM*
1	16.47361380	16.47361380	16.47361380
2	48.69674135	48.69674135	48.69674135
3	97.02096924	97.02096916	97.02096921
4	161.45155447	161.45151518	161.45150000

Table 2. Comparison of normalized buckling loads (PL^2 / EI_0) for $aL = 0.1$

Mode	*ADM*	*VIM*	*HPM*
1	13.35006457	13.35006457	13.35006457
2	39.47004813	39.47004813	39.47004813
3	78.64155457	78.64155458	78.64155466
4	130.86858532	130.86856343	130.86853842

Table 3. Comparison of normalized buckling loads (PL^2 / EI_0) for $aL = 0.2$

The buckling mode shapes of the rod for the first four modes are depicted in between Figs.5-9. To demonstrate the effect of variable cross-section in the results, a comparison is made with normalized mode shapes for a uniform rod which are given in Fig.4. Constant flexural rigidity is defined as $aL = 0.1$ in these figures.

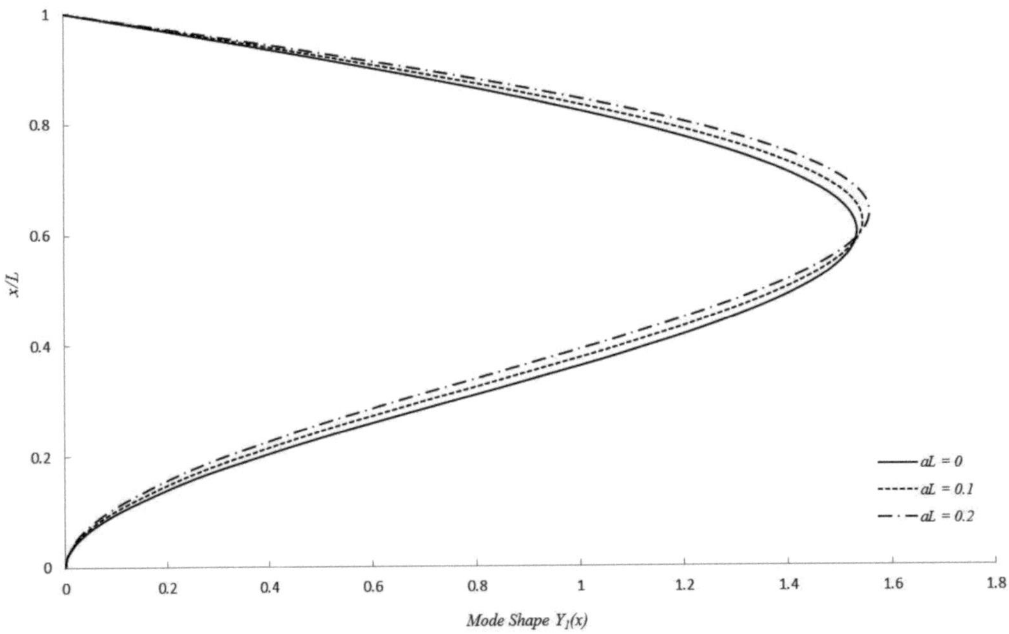

Figure 6. Comparison of buckling modes for *CP* rod (Mode 1)

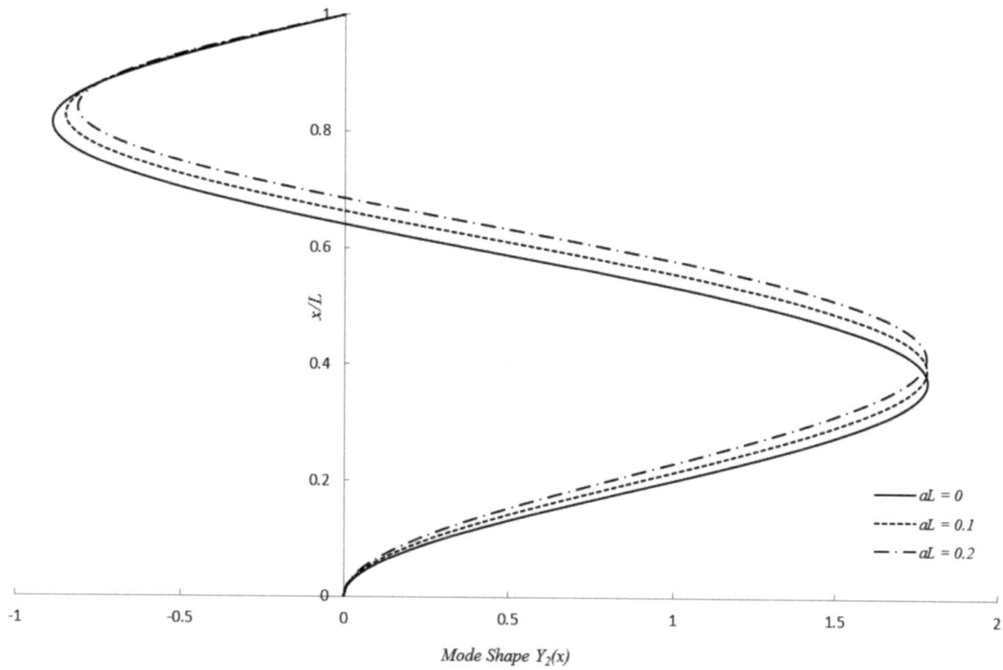

Figure 7. Comparison of buckling modes for *CP* rod (Mode 2)

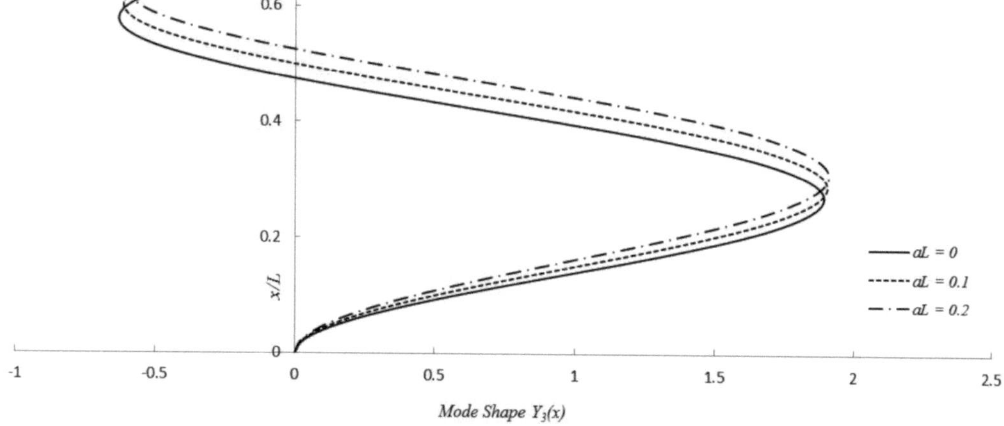

Figure 8. Comparison of buckling modes for *CP* rod (Mode 3)

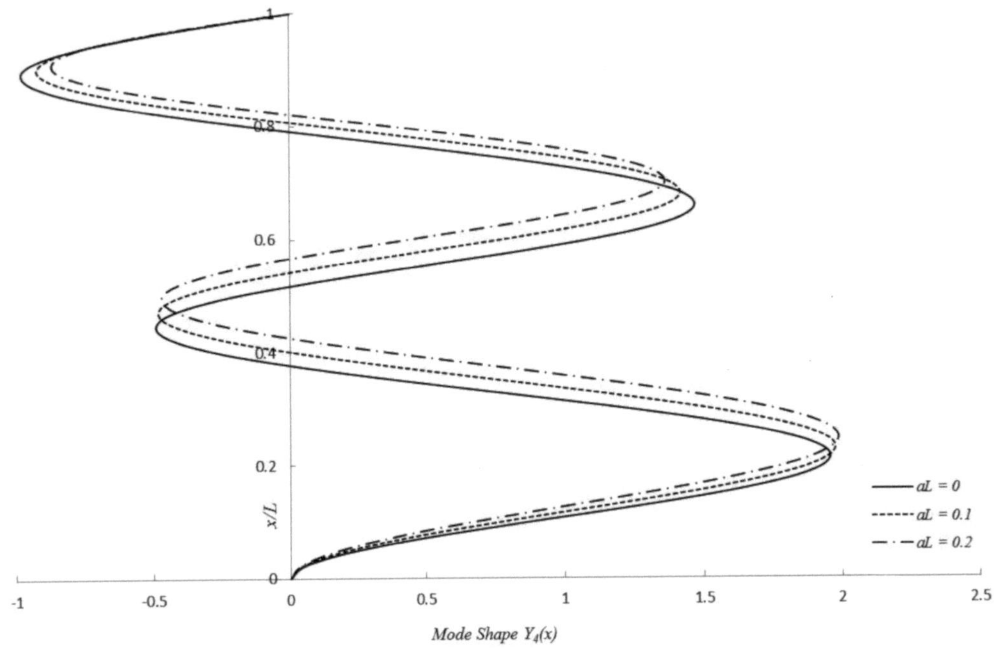

Figure 9. Comparison of buckling modes for *CP* rod (Mode 4)

5. Conclusion

In this article, some analytical approximation techniques were employed in the elastic stability analysis of Euler columns. In a variety of such methods, ADM, VIM and HPM are widely used, and hence chosen for use in the computations. Firstly, a brief theoretical knowledge was given in the text, and then all of the methods were applied to the selected cases. Since the exact values for the buckling of a uniform rod were available, the analyses were initially conducted for that case. Results showed an excellent agreement with the exact ones that all three methods were highly effective in the computation of buckling loads and corresponding mode shapes. Finally, ADM, VIM and HPM were applied to the buckling of a rod having variable cross section. To this aim, a rod with exponentially varying radius was chosen and buckling loads with their corresponding mode shapes were obtained easily.

This study has shown that ADM, VIM and HPM can be used effectively in the analysis of elastic stability problems. It is possible to construct easy-to-use algorithms with these methods which are highly accurate and computationally efficient.

Author details

Safa Bozkurt Coşkun
Kocaeli University, Faculty of Engineering, Department of Civil Engineering Kocaeli, Turkey

Baki Öztürk
Niğde University, Faculty of Engineering, Department of Civil Engineering Niğde, Turkey

6. References

[1] Vaziri H. H., and Xie. J., Buckling of columns under variably distributed axial loads, Comput. Struct., 45(3) (1992), 505-509.

[2] Dinnik, A.N., Design of columns of varying cross section, Trans. ASME 51, 1929.

[3] Karman, T.R, Biot, M.A., Mathematical Methods in Engineering, McGraw Hill, NewYork, 1940.

[4] Morley A., Critical loads for long tapering struts, Engineering, 104-295, 1917.

[5] Timoshenko, S.P., Gere, J.M., Theory of Elastic Stability, McGraw Hill, NewYork, 1961.

[6] G.J. Simitses , D. H. Hodges, Fundamentals of structural Stability, Elsevier-Butterworth-Heinemann Publishing, 2005.

[7] Iyengar NGR, Structural Stability of Columns and Plates. New York, John Wiley and Sons, 1988

[8] Wang, C.M., Wang, C.Y., Reddy, J.N., Exact Solutions for Buckling of Structural Members, CRC Press LLC, Florida, 2005.

[9] Ermopoulos, J. C., Buckling of tapered bars under stepped axial loading, J. structural Eng. ASCE, 112(6) (1986), 1346-1354.

[10] Li Q. S., exact solutions for buckling of non-uniform columns under axial concentrated and distributed loading, Eur. J. Mech. A / Solids, 20(2001), 485-500.

[11] Lee S.Y. and Kuo Y. H., Elastic stability of non-uniform columns, J. Sound Vibration, 148(1) (1991), 11-24

[12] Gere, J.M., Carter, W.O., Critical buckling loads for tapered columns, J. Struct. Eng.-ASCE, 88(1962), 1-11.

[13] Arbabi. F., Li, F., Buckling of variable cross-section columns – integral equation approach, J. Struct. Eng.-ASCE, 117(8) (1991), 2426-2441.

[14] Kerekes F., Hulsbos CL., Elastic stability of the top chord of a three-span continuous pony truss bridge , Iowa Eng. : Expt. Sta. Bull., (1954), 177.

[15] Siginer, A., Buckling of columns of variable flexural stiffness, J. Eng. Mech.-ASCE, 118(1992), 543-640.

[16] Li, Q.S., Cao, H., Li, G., Stability analysis of bars with multi-segments of varying cross section, Comput. Struct., 53 (1994), 1085-1089.

[17] Li, Q.S., Cao, H., Li, G., Stability analysis of bars with varying cross section, Int. J. Solids Struuct., 32 (1995), 3217-3228.

[18] Li, Q.S., Cao, H., Li, G.,Static and dynamic analysis of straight bars with variable cross-section, Comput. Struct., 59 (1996), 1185-1191.

[19] Sampaio Jr., J.H.B, Hundhausen, J.R.,A mathematical model and analytical solution for buckling of inclined beam columns, Appl. Math. Model., 22 (1998), 405-421.

[20] Keller, J.B.,The shape of the strongest column, Archive for Rational Mechanics and Analysis, 5 (1960), 275-285.

[21] Tadjbakhsh, I., Keller., J.B., Strongest columns and isoperimetric inequalities for eigenvalues, J. Appl. Mech. ASME, 29 (1962), 159-164.

[22] Taylor, J.E.,The strongest column – an energy approach, J. Appl. Mech. ASME, 34 (1967), 486-487.

[23] Coskun, S.B., Atay, M.T., Determination of critical buckling load for elastic columns of constant and variable cross-sections using variational iteration method, , Computers and Mathematics with Applications, 58 (2009), 2260-2266.

[24] Atay, M.T., Coskun, S.B., Elastic stability of Euler columns with a continuous elastic restraint using variational iteration method, Computers and Mathematics with Applications, 58 (2009), 2528-2534.

[25] Coskun, S.B., Determination of critical buckling loads for Euler columns of variable flexural stiffness with a continuous elastic restraint using Homotopy Perturbation Method, Int. Journal Nonlinear Sci. and Numer. Simulation , 10(2) (2009), 191-197.

[26] Coskun, S.B., Analysis of Tilt-Buckling of Euler Columns with Varying Flexural Stiffness Using Homotopy Perturbation Method, Mathematical Modelling and Analysis, 15(3) (2010), 275-286.

[27] Pinarbasi, S, Stability analysis of nonuniform rectangular beams using homotopy perturbation method, Mathematical Problems in Engineering, 2012 (2012), Article ID.197483.

[28] Safa Bozkurt Coşkun, Mehmet Tarik Atay and Baki Öztürk, Transverse Vibration Analysis of Euler-Bernoulli Beams Using Analytical Approximate Techniques, Advances in Vibration Analysis Research, Dr. Farzad Ebrahimi (Ed.), ISBN: 978-953-307-209-8, InTech, 2011.

[29] Ozturk, B., Coskun, S.B., The Homotopy Perturbation Method for free vibration analysis of beam on elastic foundation, Structural Engineering and Mechanics, 37(4) (2011), 415-425.

[30] Adomian, G., Solving Frontier Problems of Physics: The Decomposition Method, Kluwer, Boston, MA, 1994.

[31] Adomian, G., A review of the decomposition method and some recent results for nonlinear equation, Math. Comput. Modell., 13(7) (1992), 17-43.

[32] He, J.H., Variational iteration method: a kind of nonlinear analytical technique, Int. J. Nonlin. Mech., 34 (1999), 699-708.

[33] He, J.H., A coupling method of a homotopy technique and a perturbation technique for non-linear problems, Int. J. Nonlin. Mech., 35 (2000), 37-43.

[34] He, J.H., An elemantary introduction to the homotopy perturbation method, Computers and Mathematics with Applications, 57 (2009), 410-412.

[35] He, J.H., New interpretation of homotopy perturbation method, International Journal of Modern Physics B, 20 (2006), 2561-2568.

[36] He, J.H., The homotopy perturbation method for solving boundary problems, Phys. Lett. A, 350 (2006), 87-88.